REBECCA WILLIS

TOO HOT TO HANDLE?

The Democratic Challenge of Climate Change

BRISTOL
UNIVERSITY
PRESS

First published in Great Britain in 2020 by

Bristol University Press
1-9 Old Park Hill
Bristol
BS2 8BB
UK
t: +44 (0)117 954 5940
www.bristoluniversitypress.co.uk

North America office:
c/o The University of Chicago Press
1427 East 60th Street
Chicago, IL 60637, USA
t: +1 773 702 7700
f: +1 773-702-9756
sales@press.uchicago.edu
www.press.uchicago.edu

© Bristol University Press 2020

British Library Cataloguing in Publication Data
A catalogue record for this book is available from the British Library

Library of Congress Cataloging-in-Publication Data
A catalog record for this book has been requested

ISBN 978-1-5292-0602-9 paperback
ISBN 978-1-5292-0604-3 ePub
ISBN 978-1-5292-0603-6 ePdf

Cover design by blu inc
Front cover image: Alamy
Printed and bound in the UK by CMP, Poole
Bristol University Press uses environmentally responsible
print partners

Contents

Acknowledgements

I want to dedicate this book to two people who helped me immensely, in different roles but with similar humour and generosity, and who are no longer here to see the finished product: my original supervisor, John Urry (1946–2016), and my father, Dave Willis (1939–2014). I think of them often.

A huge thank you to colleagues at Lancaster: Nigel Clark and Vicky Singleton, for their encouragement, insight and humour as research supervisors; also thanks to Carmen Dayrell, Mike Berners-Lee, Nils Markusson, David Tyfield, Andy Jarvis, Jess Phoenix, Cath Hill, Andy Yuille and Duncan McLaren, who have helped me along the way.

Thanks to all at Green Alliance (past and present), particularly Matthew Spencer, Tamsin Cooper, Shaun Spiers, Karen Crane, Jo Rogers, Costanza Poggi, Alistair Harper, Frieda Metternich, Gwen Buck, Amy Mount, Melissa Petersen and Paul McNamee. A particular thanks to all those who took part in the focus group and interviews, and to the MPs' staff who were so good-humoured when I pestered.

I'm very grateful to Alice Bell, Jonathon Porritt, Keith Alexander, Tom Chance, Peter Lipman, Jon Turney and Paul Allen for comments on drafts. Thanks, too, to Claire Mellier-Wilson for thoughts on the French Assembly, to the editors and anonymous reviewers of the book proposal, and my published papers. They helped me to strengthen my arguments and dampen my ego.

Thanks to friends and colleagues, particularly Andy Dobson, Kate Rawles, Chris Loynes, Anne Power, John Hills, Lucy Stone, Pete Bryant and many others, for amazing conversations, including some brilliant chats on Twitter. Thanks to Natalie and the Steed family for the London sofa and friendship; and to Lucy Gibson, my office mate who's seen me through thick and thin.

Lastly, my family: Chris, Matthew, Katie, Jenny, William, Lana, Daisy, George and Jane, the kindest and most supportive family anyone could hope for; and my sons Sam and Jesse, who constantly remind me of the important things in life (football, apparently) and definitely ask the best questions.

ONE

Introduction: Democracy on Hold?

It's 16 September, 2019. Joanna Sustento stands alone, protesting outside the Philippines headquarters of oil company Shell. Her handwritten sign reads, simply, 'climate justice'. Joanna's parents and brother died in Typhoon Haiyan. That same day, the UK's former government chief scientist, Sir David King, tells the BBC that extreme storms are on the rise. Removing his cloak of scientific objectivity, he says "it's appropriate to be scared". Later, my teenage son, prompted by something he has read on my laptop screen, asks me what will happen if the earth heats up by 3°C. I think about what it could be like to live in a world that is three degrees warmer, the immense damage to people's lives, to human societies and to the natural world. I find that I can hardly bring myself to speak. I know it's cowardly, but I give him the quickest answer I can get away with, and change the subject. This is the climate crisis.

Faced with a problem of these proportions, the earth scientist James Lovelock has grown frustrated with the lack of response from politicians and the public. After many decades of research and advocacy on climate change, Lovelock no longer believes that elected governments are capable of taking the planetary perspective that he sees as necessary to tackle climate change. He calls for drastic measures. Likening the fight against climate change to a major world war, he points out that wars often involve the suspension of some democratic freedoms. "Climate change may be an issue as severe as a war," he has said. "It may be necessary to put democracy on hold for a while" (quoted in Hickman, 2010).

Lovelock's explicit abandonment of democracy is unusual. But it is not unusual at all for scientists, economists and other experts on climate to express deep frustration with politics and, by extension, democracy. Nicholas Stern, economist and author of the highly influential Stern Review, was speaking for many when he said 'while it is clear that it is still technically feasible to limit warming … we will not succeed without strong political will and leadership' (Stern, 2018). Fronting a climate campaign for *The Guardian*, the artist Antony Gormley showed similar exasperation, saying that politicians 'are just not capable of long term thinking' (Stern, 2018; Brown, 2015).

A growing frustration with the political system has led climate scientists to step into the breach, putting forward their own suggestions for systems of governance. In 2009, the same year that Lovelock called for the suspension of democracy, a high-profile group of scientists led by Johann Rockström used the journal *Nature* to launch their plea for a system of 'planetary stewardship' (Rockström et al, 2009). Following this approach, the first step in managing human societies is to synthesize scientific evidence and work out a 'safe operating space for humanity'. This establishes the limits that cannot be passed if the planet is to continue to provide its essential services, like provision of food, water and a stable climate. I will return to Rockström's proposals for 'planetary stewardship' in Chapter Two. For now, though, it's worth noting that, while these do not explicitly rule out democratic decision-making, the proposals implicitly suggest that democracy should take place within limits set by scientists, not politicians.

I have worked for many years on climate and energy issues, and this call for expert-led decisions is something that I see constantly. Every day, there are new reports and research saying what needs to be done. They range from the obvious and everyday, like the need for a comprehensive shift to renewable energy, to the speculative and, as yet, unproven – like technologies that promise to suck carbon out of the atmosphere. There is no shortage of experts offering solutions. And yet I see far less attention paid to the crucial question of how, as a human society, we should debate, agree and carry out plans to tackle climate change.

I don't want to accuse all climate advocates of being anti-democratic. Unlike Lovelock, a well-known contrarian, they are not suggesting "putting democracy on hold for a while" (quoted in Hickman, 2010) – at least, not explicitly. But their approach raises a fundamental question: when experts put forward their proposed solutions, under whose authority are they acting? Where does their mandate come from? They seem to be implying that democracy is all very well, as long as it takes place within boundaries already set by experts. Voters need to know their place.

In this book, I will argue that this approach is fundamentally misguided. The case I will lay out is that it is both possible and necessary to find democratic solutions to climate change. Why? Because, at the risk of stating the obvious, just because something *should* happen doesn't mean it *will*. You could call this the 'world peace' problem: we all want it, but wishing hard won't make it happen. It is also morally problematic to assume that experts always know best. Scientists may have amassed evidence of the problem, but are they best placed to dictate solutions to that problem? Economists may be good at crunching the numbers, but should we leave it up to them to decide who gains – and who loses?

Why make the assumption that citizens and voters need to be bypassed? Could we not instead assume that, if given evidence, responsibility and a stake in the process, most people would support action to protect the planet they call home?

I would suggest that the problem is not *too much* democracy, it is *too little*. I argue that, given a meaningful opportunity to have their say, most people would support action in the fact of the climate breakdown that is unfolding in front of us. But our democracies, in their current form, are just not offering people that choice. As I will show, politicians are not openly discussing the climate crisis. People have little opportunity to learn about climate action and shape political responses. Powerful interests who have much to lose, like fossil fuel companies, have too much influence.

That's why people have been taking to the streets. There have been protests about climate change before. But 2019 was different. Inspired by Swedish teenager Greta Thunberg, millions of young people around the world have walked out of school to protest at the damage being done to their futures. A 'global climate strike' in September 2019 saw millions leaving their schools and workplaces to take a stand. The protest group Extinction Rebellion has burst onto the scene, shutting down large parts of cities in the UK and round the world, in several 'rebellions', and gaining considerable public support for their action. In the US, the youth-led Sunrise Movement has been instrumental in raising climate ambition in politics. And politicians are responding. In many countries, recent months have seen unprecedented levels of attention on climate issues. New political possibilities are opening up.

And that's why, in this book, I argue that more democracy, not less, is needed to tackle the climate crisis.

There are three parts to this argument.

First, this crisis is like no other, in its extent, urgency and impact on human society (not to mention all the other species living their lives alongside ours).

Second, our existing political system needs to be understood as embedded within a society that has, for many hundreds of years, taken two things for granted: a stable climate, and plentiful fossil fuels. Neither of these can be taken for granted any more. Change is upon us, whether we like it or not.

Third, the best solutions come about if we see climate change as a collective problem, to be tackled through better democracy, not less democracy. Once we can do this, we begin to identify ways forward that can build public support for the challenge ahead.

In the course of the chapters ahead, I will develop this argument, using evidence from my own and others' academic research, and from my experience on the frontline of climate advocacy. But first, I want to set out where we stand at the moment.

We are in a difficult situation. A huge gap has opened up between the scientific consensus – which states very clearly that urgent action is needed – and the social consensus, which is conflicted, to say the least. While opinion polls report high levels of concern about climate change, most people in most countries do not see it as an immediate, pressing problem. Given this, it is no wonder that scientists and others are trying to dictate what should happen next. Let's take a closer look at this gap between the urgency of the science and the lethargy of the political and social response.

The scientific consensus: urgent action is needed

This is not a book about the science of climate change. But it is important to take stock of the evidence. The best summary I have seen recently is the opening sentence in David Wallace-Wells' meticulous and terrifying book, The Uninhabitable Earth: 'It is worse, much worse, than you think' (Wallace-Wells, 2019).

The reports of the Intergovernmental Panel on Climate Change (IPCC) may not share Wallace-Wells' lyrical writing style, but they agree (Intergovernmental Panel on Climate Change, 2018). The IPCC, established in 1988, acts as a sort of clearing house for climate science: scientists submit published papers, which are considered by a series of thematic working groups, and amalgamated into reports offering a summary of the state of knowledge. It is not without its critics, many of whom say that it is overly cautious and conservative. The way in which the structures and committees of the IPCC work, and particularly how they deal with areas of scientific uncertainty, may mean that the IPCC underplays the potential for abrupt or radical shifts in climate. The summary that the IPCC produces has to be agreed by government representatives, meaning again that it probably underplays the severity of climate science (Demeritt, 2001; Wynne, 2010). Yet even the relatively cautious account of the IPCC makes for terrifying reading.

The basic fact of climate change – that emissions of carbon dioxide and other greenhouse gases build up in the atmosphere, acting like a blanket and trapping warmth from the sun – has been understood for more than a century. Since then, scientists have worked to develop evermore sophisticated accounts of the links between emissions of greenhouse gases, average temperature rises, and resulting climate impacts.

Most recently, in October 2018, the IPCC issued a 'special report' which made a strong case for limiting average global temperature rises to 1.5°C (Intergovernmental Panel on Climate Change, 2018). The earth has already warmed by around 1°C since the pre-industrial era, so time is short. The greater the warming, the more catastrophic the impacts, including sea-level rise, reductions in crop yields, eradication of many plant and animal species and entire ecosystems, and extreme weather events like droughts, floods and wildfires. As the earth's temperature creeps higher, there are increased risks of what the IPCC ominously calls 'large-scale singular events', such as disintegration of ice sheets and melting of permafrost, which in turn have far-reaching implications for climate systems.

More speculative, but no less significant, are the impacts of these changes on human economies and societies. How will people cope in a world with less food, more extreme heat, floods and droughts? What will happen to economic and political systems, to the way that societies work and communities live together? Some parts of the planet will suffer much more severely from direct impacts like sea-level rise and crop failure. The inequality of climate impacts is gut-wrenching.

But in our hyper-connected global economy, no one will be immune to the effects.

Climate experts deal in statistics and models. This is essential: they are the tools to understanding a global, systemic crisis. But numbers – be they degrees of warming, metres of sea-level rise, categories of storm or even numbers of deaths – do not tell the human story. Human stories like that of Joanna Sustento, whose parents died in Typhoon Haiyan. Like Paul Weingartner, who only just managed to escape from the Californian wildfires in November 2018: "There are not sufficient words to describe the devastation I saw" (Weingartner, 2018). Like Abdilal Yassen, a seventy-year-old pastoralist in East Africa, one of ten million people in the region facing severe hunger: 'This drought is leaving nothing behind. In previous droughts, we used to lose some animals, but we would always have food and water. But this is different. It is "sweeping away" animals and people' (Amladi, 2017). Like the farmers who will see crop yields reduced as the earth warms, and like everyone who will have to pay more for food – if they can afford it (Schleussner et al, 2018). And like the living stories of other creatures, the white-beaked dolphins, the puffins and the leatherback turtles, three among thousands of species whose habitats, and therefore whose lives, are threatened (Díaz et al, 2019).

In short, it is worse, much worse, than you think.

The IPCC's 2018 report put a figure on how much time we have to avoid the most catastrophic impacts: global CO^2 emissions must 'start to decline well before 2030' (Intergovernmental Panel on Climate Change, 2018). This statement has given rise to the oft-repeated line that we have twelve years, now eleven or ten depending on when you are reading this, to act to avoid the worst. This line is a simplification, but not an exaggeration.

The IPCC is also clear about what needs to happen: 'rapid and far-reaching transitions in energy, land, urban and infrastructure (including transport and buildings), and industrial systems ... deep emissions reductions in all sectors, a wide portfolio of mitigation options and a significant upscaling of investments in those options'. In later chapters of this book, I will say much more about what can be done.

There is, then, both expert consensus about the causes and effects of climate change and expert consensus about what needs to happen to stem the worst effects.

The social consensus: not the most pressing problem

Since you have picked up this book and read this far, you are almost certainly an outlier in terms of your attitude to climate change. The scientific consensus is clear, as I have set out: urgent action is needed. Yet the social consensus is equally clear: for most people, in most places, climate change is not seen as the most pressing problem.

I want to stress this because, if you are concerned enough about climate change to read books about it, you will find it hard to imagine that others don't think like you. Whenever I give talks about my research into politicians' understandings of climate change in the UK, many people want to question one of my central research findings: that there has been very little political pressure, from the electorate or elsewhere, to act on climate change. I often find that people start to question my methodology, the sample of MPs I've spoken to, the polling data I've used, and so on, to try to make the case that people are making a more of a fuss about climate than my work suggests. They might also argue that this problem is peculiar to the UK – that in other countries, there is more pressure to act. Yet as I discuss in Chapter Four, my research is confirmed by many other studies. While generalized concern about climate change is high, its perceived relevance to people's lives remains low.

Among the politicians I've talked to, there is a striking consistency. As one MP said to me, "I've knocked on hundreds, literally thousands of doors, and had tens of thousands of conversations with voters … and I just don't have conversations about climate change". Nearly every interviewee said something similar, such as "I can't remember the last time I was asked about climate change. It's very rare to be asked about it", or another, who said: "When you go around with your clipboard asking what are your top priorities, you always know it's health, economy, education, crime, stuff like that, and environment always comes very [low] down."

A number of interviewees did, however, suggest that there was an exception to this pattern. They reported that there were people within their constituency who raised climate change, normally as part of a broader set of issues concerning the environment and social justice. One MP described this group as "articulate, affluent people who have perhaps a particular type of worldview". Another said that his patch included "Guardian-reading intelligentsia, who are engaged … we know from the emails they send that climate change is one of the issues of concern". I think it's probably the case that most of us who work on climate issues, and who read books about climate change, are part of this group. We need to be careful not to forget that we are the exception, not the rule.

This same MP, though, made what he saw as a very important distinction between this small group and "the other eighty-seven thousand people, and particularly working-class people, who are not going to be engaged in the issues, but are concerned about whether their kids can get to school or whether the hospital is operating".

In that last response lies the core of the issue. Climate change is not (yet) part of the everyday batch of issues which pre-occupy people or which they want their politicians to engage on. That is not to say people don't care or aren't worried. The fact that there has been very little political pressure to act on climate change does not necessarily mean that there is no mandate for action, or that it is not possible to craft a politically popular agenda for climate action. And things are changing rapidly: the recent upswell in protest, public concern and pol-itical attention to climate shows that change is possible. The UK's election campaign in 2019 certainly had more discussion of climate than any other election I've known, including the first ever televised debate between party leaders on climate change. For once, parties were competing over who had the best climate policies.

What lies ahead?

This book is in two parts. The next four chapters look at the current state of climate politics, and the final three offer some ways forward.

In Chapter Two, I look at what it might mean to govern climate change: to muster a collective, political response to the challenge. I first ask whether it is realistic to imagine that we can steer the planet, like astronauts on a spaceship. I contrast these 'spaceship earth' ideals with existing attempts to govern climate change. On the international stage, these efforts began in 1997 with the global Kyoto Protocol, and have continued, with varying degrees of success, ever since, culminating in the Paris Agreement of 2015. I also look at how different countries, including Australia, the US, the UK, China, Ethiopia and the Maldives, have addressed the issue. This quick survey shows that responses to climate change depend heavily on each country's social and political understandings – in short, responding to climate change is political. It also shows that no country has, yet, done enough to claim that they have implemented a comprehensive or adequate response to the scale of the challenge.

The reasons why current national strategies fall short of the mark are discussed in Chapter Three. In this chapter, I highlight the extent to which society in richer countries is dependent on constant energy input, which is mostly derived from burning fossil fuels. Citizens in these countries are so accustomed to high-carbon systems that it is very difficult for them, or their political leaders, to envisage a low-carbon society. The intransigence of material and social systems, and the lobbying reach of the fossil carbon industry, fights against change.

In Chapter Four, I describe the dual reality in which we now live: acknowledging the enormity of climate change, while life goes on around us. The way in which both politicians and people respond to climate issues depends on how they navigate this dual reality. Although concern about climate change is rising around the globe, it is accompanied by what might be called 'societal denial' – a reluctance to think through the implications of this for political or social life.

Chapter Five looks at what has been done so far to address climate change. This chapter concludes that there are grounds for optimism in solutions put forward, from the boom in renewable energy to cities implementing zero-carbon development strategies. But so far, climate action has been limited by two factors: first, a 'feelgood fallacy', in which strategies are focused on encouraging new solutions, rather than attempts to halt damage – like reducing demand for high-carbon transport, or curbing consumption levels; second, 'stealth strategies', which assume that experts know best and try to impose solutions on an unthinking public.

In Chapter Six, I turn to the question of what can be done. What would happen if we started to see democracy as part of the response to climate change, rather than a hurdle to overcome? First, I look at how politicians think through their role as representatives, and their relationship with the electorate. Based on this understanding, three shifts are proposed: a more deliberative model of democracy, in which politicians, citizens and experts debate and collaborate on climate strategies; a much clearer story of transformation and transition, away from a high-carbon society and toward a post-carbon future; and last, an acknowledgement that climate action is about more than evidence and technical fixes – it is an appeal to the heart as well as the head.

In Chapter Seven, I take these democratic principles and use them to put forward a strategy for the climate emergency. Such a plan will, by its nature, be a constant negotiation between citizen and state – a social contract. If it is a meaningful response to the challenge, it will include both an upfront acknow-ledgement of the climate crisis and a story of transformation, accompanied by careful measurement and management. It will make clear that climate action is about stopping some things, like fossil fuel extraction, as well as doing more of other things, like renewable energy. It will cede power, and responsibility, to local areas, allowing politicians to craft local strategies that are meaningful and tangible to local people. And it will be open about the international implications of individual country strat-egies, supporting climate action elsewhere. I set out ten basic principles that should underwrite all national climate strategies.

Last, in Chapter Eight, I turn to the personal. What does it mean to be a good climate citizen? We are bombarded by advice about how to shop and how to eat, but what can each person do to contribute to better climate politics? I finish on an unashamedly optimistic note. I don't see what alternative there is.

The background to this book: research theory and method

What gives me the authority to argue this? After all, given that my case centres on the problems of an expert-led approach, I could be accused of undermining my own status and authority. In a way, this is exactly what I want to do. I want to encourage everyone who has a professional role related to climate change – whether in an environmental organization, the political world or the business or research community – to step back and think not just about what we judge needs to be done, but also how. Through what social processes, and with whose contribution and engagement, do we believe change can be achieved?

That is not to say that I am somehow anti-expert, that I reject the scientific consensus or the policy proposals that experts put forward. Rather, I want to argue that expertise is an essential resource, to be deployed in good faith, as part of a process of engagement and social dialogue – not as an imposition from above. With this caveat, let me set out the research process and evidence base that forms the bedrock of this book.

This book stems from twenty years of practical experience of working on climate politics, and from four years of academic research at Lancaster University. I have worked on climate politics since 1997: first, at the European Parliament, as an adviser to two politicians; then, at the think tank Green Alliance, at the point when countries were just starting to consider what climate change would mean for domestic politics. From 2004 until 2011, I had an inside track – as a vice chair of the Sustainable Development Commission, a government-sponsored advisory body. We had the interesting and definitely challenging job of advising the Prime Minister, the UK government and the leaders of the Scottish, Welsh and Northern Irish legislatures on a broad set of environmental and social issues, including climate change.

In 2009, just after the UK had introduced its landmark Climate Change Act (which I discuss in Chapter Two), I had an idea that would come to dominate my working life for ten years. Frustrated by what I saw as the marginal position of climate change within political life, I wanted to make sure that the next generation of politicians was equipped with all the necessary evidence and skills to make informed decisions on climate. I asked around, and discovered that Members of Parliament are offered very little in the way of training or support for their new, often daunting, role. So, working with Green Alliance and with the backing of some far-sighted charitable foundations, I set up the Climate Leadership Programme.

Through the Programme, we offered parliamentary candidates the chance to learn more about climate science and politics, before they had even been elected. We also worked with new MPs, to offer support and advice. We ran workshops for each main political party, introducing politicians to the science, policy and politics of climate change. We asked leading climate scientists to join us to answer the MPs' questions. Business people came to talk about how they were thinking about the likely impact of climate change on their companies. Climate activists explained why they campaigned on the issue. At the end of each workshop, we invited senior politicians to come and hear from the candidates about what they had learned, and discuss the implications of climate change for each party's political strategy.

The workshops were undoubtedly a success. We trained over 100 politicians over two electoral terms, and the programme continues. It was clear from the discussions they had during the workshops that they understood the significance of climate change. But the more time I spent on the work, the more I was nagged by a fundamental question. We offered the politicians the chance to learn, and to think. What happened when they walked out the door? How did they take this learning into their work as a politician? To put it another way – suppose each politician had left our workshop convinced of the case for action on climate, how would that change the way they approached their job? The more we knew about this, the more we would know about how to support politicians, to challenge them when necessary, and from there to build a more constructive political environment for climate action.

The more I thought about these questions, the more intrigued I became. I realized then that the best way to find answers might be to turn to academic research methods. A conversation with some very patient friends at Lancaster University turned into a full-blown research project, and it is the insights from this project, over four years, as well as from my working life, which form the basis of this book.

Probably because I am a latecomer to academia, the theoretical basis for this book is wide-ranging, or 'pleasingly eclectic', as a kindly colleague once remarked. It draws on sociology, political science and theory, science studies and environmental governance, in the main.

I discovered that there had been very few previous studies similar to the one I designed, investigating politicians' personal responses to climate change (I could only find one, in fact, looking at Australian politicians (Fielding et al, 2012)). But there was some fascinating work in sociology and political science, on the more general question of how politicians define their identity and approach their work (Childs, 2004; Crewe, 2015; Fenno Jr., 1977; Lovenduski, 2012; Malley, 2012; McKay, 2011; Norton, 2012). I discuss this in Chapter Four. More generally, there is a vast and fascinating body of work from the field of science and technology studies, examining how people understand scientific evidence. This was really useful to me in thinking about what politicians, and the rest of us, do with the scientific information that is thrown at us (see for example Jasanoff, 2010; McNeil, 2013; Wilsdon and Willis, 2004; Wynne, 2010). There is also a huge amount of careful scholarship on climate governance, from the planetary scale – how to govern a planet, as I discuss in Chapter Two – down to detail of how to design a carbon tax. Last, a growing body of scholarship questions the fundamentals – asking the big questions about whether tackling climate change is compatible with a capitalist economy or the pursuit of growth (Jackson, 2017).

To find answers to my specific questions about how politicians understand climate change, I carried out three specific pieces of field research. First, I set out to examine how Members of Parliament spoke about climate change in the public forum of the House of Commons, using a method called corpus analysis. This method uses statistical techniques to examine the tone and structure of language, and build up a picture of politicians' understandings. Second, I brought together a group of people who work with environment and development organizations, including the World Wildlife Fund, Christian Aid and Greenpeace, for example. I used focus group methods to ask them about their experience of working on climate politics, and the ways in which they interacted with politicians.

Last, I set out to get the views of the politicians themselves. Over two years, I interviewed 23 Members of the UK

Table 1: Interviewees' background and experience

Gender	14 male, 9 female (gender balance of current Parliament is 71% male)
Party affiliation	8 Conservative, 9 Labour, 4 Liberal Democrat, 2 other
Time served as MP	Between 1 and 23 years' work as an MP; mean = 8.6 years
Current status	12 sitting MPs; 11 former MPs, who left office between 2010 and 2017
Seniority	9 interviewees had served in government; 4 had served on the opposition frontbench. 10 were backbenchers, with experience on Select Committees
Record on climate change issues	7 with a strong record of activity on climate change issues (assessed through speeches in Westminster and elsewhere); 11 with some activity; 5 with little or no activity

Note: Participants were not asked for additional demographic data, eg age or ethnicity.

Parliament. I aimed for as representative a sample as possible, with politicians from a range of political backgrounds, age and experience. Table 1 sets this out. Rather than taking a question-naire approach, I conducted the interviews as a conversation, or narrative (Riessman, 2008). I had the same basic structure, but was guided by the interviewees, and followed up useful lines of discussion as they arose. I asked MPs to reflect on the way they work in general, and the influences and pressures upon them, before asking questions specifically about climate change, including their views on how it is discussed in Parliament, as well as their own viewpoint.

I offered the politicians anonymity, and in return, I was surprised and grateful for their openness. Many displayed a disarming frankness about their professional and personal struggles, often at odds with outsiders' perceptions of their role.

A series of papers in academic journals offer more detail on the methods used, and the specific findings of each piece of fieldwork (Willis, 2018a, 2018b, 2017). In this book, though, I draw more freely on this set of findings, my own career, and evidence from other countries, to discuss the issue of climate change and democracy more generally.

As I have described, the politicians I spoke to were not sure what level of support there would be from their voters for action to tackle climate change. So to follow up this research, in 2019, I worked with Green Alliance and a public research specialist, Britain Thinks, to carry out some research workshops, called 'citizens' juries'. In two contrasting places, Cardiff and Penrith in the North of England, we brought local politicians together with small groups of constituents. The participants spent a day learning about climate science and impacts, discussing different strategies for tackling it, and offering their recommendations for government and politicians (Buck, 2019). I write about these findings in Chapters Seven and Eight.

The fieldwork for this research only covered the United Kingdom. However, throughout this book I make comparisons across countries. While UK politics has many fascinating quirks and specificities, the work I have done offers more general answers to questions about the relationship between democracy and climate change.

Finally, a word about 'we'. In this book, I often say what 'we' think, or what 'we' need to do. This is partly stylistic – any editor worth their salt will tell you to avoid using the passive voice wherever possible. It also serves to emphasize that climate change is a collective problem, affecting all of humanity, and that each of us is both part of the problem and potentially part of the solution. There are, though, risks in too liberal a use of 'we'. It might be taken to suggest that each of us is equally responsible, even though the carbon footprint of individuals in low-income countries is thirty-five times lower than individuals in high-income countries (World Bank, 2019). The richest ten per cent of people in the world are responsible for half of all emissions (Oxfam, 2015). Talking about the universalized 'we' of humanity might mask the power and responsibility of individual companies and countries, an issue which I discuss in Chapter Three. And it risks accusations of speaking for humanity as a whole – what the philosopher Donna Haraway calls 'the god trick of seeing everything from nowhere' (Haraway, 1988, p 581).

So let me be clear about where I stand. I am writing as me – a relatively wealthy, privileged individual in a rich country. I write from my own research and experience, focused on the UK, but in a global context. As I hope I will make clear in the chapters that follow, the climate crisis is as much about power and politics as it is individual actions. We are in this together, but each of us has vastly different impacts, different choices and limitations, and – most importantly – different ways of contributing to climate action.

TWO

A Minute to Midnight: Governing the Planet

It's 1982, and I'm ten years old. A geography lesson in my final year of primary school. We've just found out how old the earth is, and we're compressing those four and a half billion years of planetary history into a single day. A giant timeline stretches round all four walls of the classroom, with a metre for each hour of the twenty-four hours of that day. We draw Earth's history onto the timeline, starting with the planet forming, at midnight. Until three in the morning, or just over a billion years in real time, meteors bombard Earth. At four am, life begins, but we have to wait another fourteen hours – until six in the evening – for sexual reproduction.

Things speed up after that. The last metre or two of the timeline is crammed with excitement. By ten o'clock at night, the vast swamps that will become our reserves of oil, coal and gas are laid down. Dinosaurs appear at eleven o'clock. Then, just under a minute and a half before midnight, the first humans appear, though not yet our own species, *Homo sapiens*. All of our history as a species – cave dwellers, the Romans, the Industrial Revolution, and of course our own lives – is crammed into the final few seconds before midnight.

It is a dizzying realization for a ten-year-old. I wonder what the earth looked like for all those years, before it was shaped by plant and animal life. Those dinosaurs are suddenly near relatives. My life, my species even, seems an insignificant speck upon a vast, old planet.

I remember this lesson many years later, when I start thinking about climate change. Because it's really only in the last twelve thousand years or so, a fraction of a second on that timeline, that civilization as we know it emerged. The development of agriculture allowed people to group together in villages, towns and cities, leading to larger, more complex societies, forerunners to our own interconnected world.

It is no coincidence at all that the earth's climate has been more stable for those twelve thousand years than at any point in the long history of our planet. All those things we take for granted – farming, industry, settlements, culture and, of course, politics itself – have happened during a time of remarkable climate stability. And as we now know, this stability will not continue.

We are not very good at thinking on planetary scales. We are focused on that second before midnight. It may be a blink of an eye in planetary terms, but contained within that second is everything that is familiar to us, everything that we can under-stand and relate to. The incredible, long, turbulent history of our planet would be little more than a matter of curiosity, if we could take for granted the sort of conditions that we have enjoyed during our own, brief habitation of Earth. But we can't.

The science of climate change wasn't in the curriculum when I was at school, though the science was already well understood. The teacher could have pointed out that in those few seconds before midnight, the earth's climate has been both more stable and more accommodating to complex life than at any point in the preceding hours. And that, in a tiny fraction of a second just before midnight, as the Industrial Revolution unfolded, humans have altered the earth's climate system more quickly than the planet has ever seen before.

Spaceship Earth: can we take control of earth systems?

Johann Rockström, a straight-talking Swedish professor, has made it his life's work to explain the complex dynamic systems that make up our planet. With his colleagues, he has reviewed a huge amount of scientific evidence, to present ten aspects of earth systems, including biodiversity, the nitrogen cycle, fresh water and, of course, climate (Rockström et al, 2009). All these interconnected systems, Rockström argues, must continue to function effectively, to maintain the living conditions that humans, and other species, rely on. Rockström's work is a simplification of an incredibly complex and uncertain scientific picture. But that is the point of it. He has distilled everything you need to know about living conditions on our planet onto a single PowerPoint slide.

The aim of Rockström's work is not merely to describe how the planet functions. Once we have this understanding, he argues, we can use it to manage our impacts on the planet. For each aspect of earth systems, we can use the evidence to set a boundary, deciding at what point it would be unwise to cross this threshold. Taken together, all these limits can be seen as a set of 'planetary boundaries' which define what Rockström calls a 'safe operating space for humanity'. Transgress these boundaries and we risk undermining our planet's ability to support us.

In this way, Rockström argues, it is possible to develop a system of governance, based on scientific evidence, to manage our impact on the planet. He calls this approach 'planetary stewardship'.

The concept of planetary boundaries is an incredibly helpful way of distilling complex information into an accessible description of the life support functions that planet Earth provides. It is very hard to argue against the idea that we should keep within the boundaries that Rockström describes.

Yet there is a real difficulty with the idea of planetary stewardship. According to this view, there is a need for a global, planetary-level, science-based governance regime, run by the United Nations or a similar global institution, which all countries must agree to be bound by. In short, planetary stewardship rests on an assumption that it is both possible and desirable to govern Earth as some form of global authority. Is this really the case?

First: is it possible to govern the earth? There have certainly been attempts to do so. The United Nations was established with exactly this in mind. Countries agreed to bide by a set of internationally agreed procedures and laws, in order to manage issues that could not be handled by individual states. The record of the United Nations in areas like peacekeeping, support to refugees and, of course, global environmental challenges shows that international cooperation is possible – up to a point. Yet the many failures it has suffered during its brief existence also shows that global government is fraught with difficulties. Later in this chapter, I will describe in detail how the United Nations has tried to govern climate change. For now, it is enough just to say that there is no smooth, agreed process by which scientific evidence can be translated into international action. Yet, reading the case for 'planetary stewardship', such a process seems to be assumed.

A commonly used metaphor is that of 'Spaceship Earth' – our planet as a spacecraft gliding through the solar system, steered from a central cockpit. Using this metaphor, science provides information that is relayed to the cockpit, showing where the spaceship is heading, whether its systems are operating smoothly, whether there is enough fuel, and so on. With this information, the pilots of Spaceship Earth can maintain safe operating conditions, through adjusting the dials and levers in front of them. In this way, the pilots steer Spaceship Earth according to scientifically agreed procedures. The eccentric futurist Richard Buckminster Fuller popularised this concept as long ago as 1969. He went so far as to write an 'operating manual for Spaceship Earth'. The manual ends with a confident assertion: 'politicians can and will yield enthusiastically to the computers' safe flight-controlling capabilities in bringing all of humanity in for a happy landing' (Fuller, 1969, p 44).

It hasn't quite turned out that way. While the metaphor is beguiling, it opens up an immensely long list of questions. Who are the pilots? Who has decided that they can steer Spaceship Earth? What role is there for the rest of us mere mortals, the passengers on the spaceship? Are we allowed in the cockpit? Do we get any say over how things are done? Who designed the dials and levers, what do they measure, and what is left out? I could go on. The Dutch political scientist Maarten Hajer uses the term 'cockpitism' to describe 'the illusion that top-down steering by governments and intergovernmental organizations alone can address global problems' (Hajer et al, 2015, p 1652). A more accurate picture would acknowledge that scientific knowledge claims are not, in themselves, enough of a foundation for global cooperation. Attempts at international coordination will always be argued over; there will be power struggles within and between countries; and it is wrong to assume that we will ever agree a single strategy for managing our complex planet.

Instead, the more frightening truth is as follows. We know that human action has led to far-reaching changes to Earth's systems. But that knowledge in itself does not lead to a solution. In the words of the academic Noel Castree:

> ... humans are party to a huge and unrepeatable biophysical experiment in which we are not mere observers but a key part of the experiment itself. This is not, of course, to suggest we are somehow in control nor even pilots able to successfully steer the metaphorical (space) ship. (Castree, 2014, p 444)

It is clear, then, that it is not possible to steer the planet like a spaceship. But, even if it were, would it be desirable? Should we be striving for stronger international cooperation and global agreement about what must happen? Under whose authority are we acting?

Some commentators are extremely critical of the planetary boundaries approach. In a memorable exchange with Rockström, the technology scholar Andy Stirling states his concerns that concepts like 'earth system governance' and 'planetary management' make it seem as if the problems are technical and scientific, to be managed by scientists acting in the best interests of humanity (Stirling, 2015).

Instead, Stirling says that 'the realities of control are often far removed from the romanticised visions. What incumbent power likes to present as control, is often better understood as being about reproducing privilege' (Stirling, 2015). Focusing on top-down control, he writes, obscures the real problems, of 'entrenched industrial interests, technological infrastructures or cultures of inequality and consumption'. In other words, the notion of 'planetary stewardship' assumes that problems can be solved through top-down science-based rational governance, rather than through addressing the unequal relations of power and privilege which Stirling holds accountable for many of the environmental problems that we face.

Stirling goes further still, to suggest that top-down analysis inevitably leads to big technical solutions, such as geoengineering. Geoengineering refers to a range of suggested ways of intervening in earth systems to correct supposed imbalances in those systems – for example, putting mirrors in space or seeding clouds to reduce the amount of sunlight that reaches the earth. If you think in terms of big systems and top-down control, Stirling argues, your proposed solutions will be similarly big, technical and controlling. Such an approach is likely to reinforce existing patterns of privilege, both within countries and between countries (Scoones et al, 2015).

Whether or not you agree with Stirling's critique, it is certainly true that the concept of 'planetary boundaries' is silent on the issue of democracy. Under whose authority would governance interventions be made? Governance of earth systems is presented as a precondition of human society, something that *must* be done, given the weight of scientific evidence. The question of *how* reforms can be brought about, democratically or otherwise, is not specified. The political theorist Andrew Dobson describes this approach, stating what 'must' be done without specifying how, as attempting to play 'a card that will trump political debate and discussion' (Dobson, 2010, p 265).

I'm not accusing advocates of planetary boundaries of arguing against democracy, by stating that scientists should steer the planet in the best interests of humanity. Planetary boundaries enthusiasts more likely see their job as one of persuasion – persuading governments, parliaments and electorates alike that such steps are necessary. This may be more democratic, but, as Stirling points out, it does not leave a lot of room for discussion of conflicting visions, vested interests or power relations, as I will discuss in Chapter Five.

In the planetary boundaries concept, then, we see a clear statement of what needs to change, without any indication of how that change could be brought about. We see an admirably clear diagnosis of the problem, which leads to an unhelpfully and unrealistically simple account of the solutions. Turning to climate change in particular, we can see how the reality of 'earth system governance' has played out, since the issue was first raised at the United Nations, over thirty years ago.

Governing the climate: from Kyoto to Paris and beyond

I was in my second year at university when efforts to govern climate change began in earnest. I remember it well. The early 1990s were a time of real hope for environmental governance. A few years before, scientists at the British Antarctic Survey had measured surprisingly low concentrations of ozone at the South Pole, dubbed the 'ozone hole'. The science was clear: human-made chemicals, called chlorofluorocarbons or CFCs, were responsible. What followed was remarkable. Governments signed a landmark environmental convention, the 1987 Montreal Protocol, pledging to phase out CFCs. This was a case of multilateral action at its best. We now know how successful it has been: the hole is slowly healing (Strahan and Douglass, 2018).

There were high hopes that this successful approach could be more widely applied, and in particular, to the emerging problem of climate change. Following increasing concern from scientists, the Intergovernmental Panel on Climate Change (IPCC) had been created at the end of the 1980s, and published its first report in 1990. In response, the United Nations (UN) agreed to begin work on an international treaty on climate change. The new cause had some surprising allies. In 1989, the UK Prime Minister Margaret Thatcher made an impassioned speech to the UN's General Assembly. The first forty years of the UN's history, she argued, had been concerned with what she called the "accustomed perils" of international diplomacy. The next forty years would be marked instead by "another insidious danger ... the prospect of irretrievable damage to the atmosphere, to the oceans, to earth itself" (Thatcher, 1989).

Reading Thatcher's speech today, I am surprised by how contemporary it sounds. The language may be a bit old-fashioned. Her phrase "man-induced climate change" seems to absolve half the population of any responsibility at all. But the scientific case she sets out is remarkably similar to any modern account. Nor does she see it just as a problem for the future – she gives examples of effects already felt, like thinner sea ice. Since then, scientists have refined their models, built up vast quantities of evidence, and increased the certainty of predictions – but the basic scientific case remains unchanged.

And so the international community threw themselves into action. 1992 saw the birth of the climate treaty that still holds today: the United Nations Framework Convention on Climate Change, whose stated aim is to 'stabilize greenhouse gas concentrations in the atmosphere at a level that would prevent dangerous anthropogenic interference with the climate system' (UN, 1992). Just as with ozone depletion, it seemed that climate change, too, could be managed through careful international cooperation.

And yet already, back in the early 1990s, it was becoming clear that this was a problem of a different sort altogether. From my vantage point in Cambridge's University Library, I wrote an essay about why climate change was a more difficult problem to solve than ozone depletion. Whereas ozone-depleting chemicals could relatively easily be substituted for safer substances, I wrote, the same could not be said for greenhouse gases. Whereas there were only a few companies producing ozone-depleting chemicals (these, incidentally, lobbied heavily against the Montreal Protocol), greenhouse gases were an unintended by-product of nearly all industrial processes, transportation and energy generation. Plenty of companies, and indeed countries, worried that they had much to lose from restrictions imposed for the sake of a distant threat to the climate. Or so my essay argued.

I would like to claim that I, precocious as I no doubt was in my undergraduate years, was uniquely able to predict the problems ahead. But my undergraduate ramblings were really just a summary of a dawning realization shared by many: ozone was easy, climate would be hard.

So it has proved to be. Over the intervening years, the international process has continued, in a mix of leaps and stumbles. At a meeting in the beautiful Japanese city of Kyoto, in 1997, there was a giant leap. Richer countries agreed to limit their emissions of greenhouse gases, and signed up to legally binding targets. Developing countries, whose emissions per head were far lower than the developed world, were not required to reduce emissions, but given funding and support for low-carbon industry, forest protection and other climate-improving measures. But political difficulties within and between countries meant that negotiations stalled. Another eight years would elapse before enough countries ratified the Kyoto agreement, allowing it to enter into force in 2005. Every year, negotiators from every country met to try to thrash out a way forward, but the going was tough. In 2001, President

George W Bush withdrew the US from the process altogether. He gave two reasons for this: first, the "incomplete state of scientific knowledge", and second, that the agreement would "harm our economy and hurt our workers" (Darwall, 2017).

By then, I was in London, running the environmental think tank, Green Alliance. I remember hearing that the US was pulling out of the Kyoto process just as President Bush arrived in the UK for a state visit. I saw the Presidential helicopter fly overhead, and it felt like a personal snub. There was anger, and a sense of helplessness too, that the US had taken a wrecking ball to years of negotiation – and hope. And then a few years later in 2009, came the failure of the Copenhagen climate talks. Billed as the best possible chance of developing a strong, multilateral climate agreement, the talks ended in disagreement and acrimony, wounded by the US' reluctance to participate and the shock of the global financial crisis. The only thing salvaged from the talks was a commitment to keep negotiating.

For me, the abiding image of the climate talks during these years was that of Yeb Sano, a negotiator for the Philippines government, who wept as he described the devastation caused by Typhoon Haiyan, and pleaded with fellow delegates to "stop this madness" (Memmott, 2013). These years were a brutal reminder that strong scientific evidence does not, in and of itself, provide the grounds for action, whatever people might say about planetary stewardship and Spaceship Earth. It was obvious then, and is still obvious now, that a strong multilateral agreement is badly needed. And yet those years also demonstrated with a blinding clarity that, just because an agreement is needed, it doesn't mean that it will happen.

Copenhagen was a low point. Yet it sowed the seeds for the Paris Summit of 2015, which, against the odds, did result in a global agreement to tackle dangerous climate change. What happened in the intervening years, between Copenhagen and Paris, to move to a successful agreement? One of the main shifts was, in fact, a move away from an assumption that a top-down, Spaceship Earth-type process could solve the problem.

Until Copenhagen, all the UN climate talks had, essentially, been attempts to use the international process to impose carbon reduction targets on individual countries. The negotiations had centred on so-called 'burden-sharing' – by which countries would accept limits to their carbon emissions. Under this arrangement, historically industrialized countries, including European nations, the US and others, agreed to limit their emissions. Emerging economies, including relatively prosperous nations like China and India, were not required to do so. This was one of the reasons that the US, under President Bush, took umbrage: from their point of view, they were being required, by the UN process, to share a 'burden' while other countries were escaping scot free. I am not saying that I agree with this view. Industrialized countries have emitted much higher levels of greenhouse gases, over much longer periods of time, and so have a higher degree of responsibility for climate change. As richer countries, they are also more able to invest in carbon reduction strategies. But if you are the US President, the fact that you have had limits imposed on you while other countries have not is a hard thing to sell to your electorate.

Leading up to the Paris Summit, the failed top-down approach was turned on its head. Rather than top-down burden-sharing, the proposed approach was known instead as 'pledge and review'. This works as follows: countries agree an overriding goal, to limit dangerous climate change. Then, rather than fighting over how the 'burden' is allocated, each country is free to put forward a pledge – known as a 'nationally determined contribution' or NDC. The NDC, in effect a climate action plan, is a chance for each country to set out the contribution it will make – how it will reduce its emissions of greenhouse gases, and by how much. The huge advantage of this approach is that it is up to the countries themselves to say what they will contribute to the global effort. It is not imposed through the UN, but agreed by each national government. All countries have the same responsibilities, though there are mechanisms to provide finance and support to poorer countries. Once the NDCs have been drawn up, they are reviewed – pulled together and added up, to see if the collective effort is enough to meet the scale of the challenge. If it is not, countries are then asked to step up their own efforts, again through nationally decided action. Using this process, it is hoped, countries agree the overall goal, while retaining control over how they will contribute to the global effort.

The pledge-and-review approach comes with its own risks. It relies on countries putting forward meaningful responses and agreeing to contribute to collective action. So far, the pledges that countries have put forward fall far short of the overall reductions required. It has been estimated that, if all the pledges are honoured in full, emissions levels will result in over 3°C of warming, compared to the 1.5° that has been agreed as a safe level (Fawcett et al, 2015). As a result, since 2015, the focus of the negotiations has been in developing the so-called 'Paris Rulebook', trying to agree common rules to govern national contributions.

This short history of climate diplomacy demonstrates the problems inherent in top-down approaches to governance. It is not at all clear whether the Paris system will be more effective. The success or failure of global climate management now depends on action within each country. This is what we now turn to.

Governing the climate at national level

As the Paris climate negotiations entered their second week, back in December 2015, my home town disappeared under water. Kendal is a pretty market town with a river running through it, taking rain from the Lake District fells out into Morecambe Bay. That week, the ground was already saturated, and it just kept on raining. The schools all closed (my children were delighted), the roads were impassable, and many houses were flooded. I live up a hill, so my house was fine, but it was soon full of friends and family who couldn't go home. It took quite a while for Kendal to get back on its feet. In poorer countries, with less resilient infrastructure and fewer resources, the impact of such extreme weather is not inconvenience and damage to property: it is loss of life and livelihoods.

More and more people are, like me, feeling the effects of climate change firsthand, even if they don't make the connection. Kendal's recent floods are consistent with climate forecasts for the UK. Drier summers and warmer, wetter winters are predicted, with more of the prolonged, intense rainfall which caused the devastating deluge in 2015. Yet it was hard, even for me, a seasoned climate watcher, to make the link between events on the international stage in Paris and the devastation in my back yard. Links between climate change as a global issue, and implications for each country, are hard to make.

However, since that first international agreement in Kyoto in 1997, countries have been thinking through their own contributions to climate change. I want to have a look at what has happened in a few, very different, countries – Australia, the US, the UK, China, Ethiopia and the Maldives.

You could see Australia as a microcosm of climate politics. Increasing temperatures, drought, and the devastating bush fires of summer 2019–20 have lodged climate change in the national consciousness. Yet Australia is one of the world's biggest exporters of coal and has a politically powerful fossil fuel lobby. As a result, politicians in Australia are divided – more so, probably, than in any country other than the US. Australia and the US were the only countries to withdraw from the Kyoto Protocol. There is a clear split between those on the left, who broadly accept the scientific consensus and advocate climate action, and those on the right, who question the science.

A proposed carbon tax was widely seen as one of the reasons why Prime Minister Malcolm Turnbull was brought down in 2018, with Scott Morrison, famous for his support of the coal industry, taking his place. Morrison did an abrupt about-turn, proving obstructive at international climate negotiations and withdrawing money from the Global Climate Fund, designed to help developing countries act on climate. Yet the fightback was fierce too. In November 2018, thousands of school students took to the streets, in some of the earliest of the now-global school strikes. Scott Morisson roundly condemned the protest, saying "What we want is more learning in schools and less activism". One student responded with a placard which read, 'We'll be less activist if you'll be less shit' (Guiney, 2018). Never mess with an angry teenager.

As in Australia, climate politics are deeply partisan in the US. While a Republican president withdrew the US from the Kyoto Protocol, a Democrat president, Barack Obama, was instrumental in securing the Paris Agreement. President Trump's sceptical views on climate change are well known, and a recent review concluded that he has a 'campaign to systematically walk back US federal climate policy' (Climate Action Tracker, 2019a). But California (which would be the world's fifth largest economy if it were a country) has an impressive record on climate, leading the way in many policies.

Having split the country, climate politics are now splitting the Democratic party. A crop of radical Democrats, including recently elected Congresswoman Alexandria Ocasio-Cortez and backed by the youth-led Sunrise Movement, are fighting to put climate action at the centre of Democratic politics. Their flagship proposals for a Green New Deal (discussed further in Chapter Six) are opposed by so-called 'moderate' Democrats. The 2019 race for Democratic nominations scored a further climate first, when Jay Inslee, Governor of Washington, campaigned explicitly as a climate candidate, though his support quickly faltered. These controversies, within and between parties, have put climate change on the US political agenda as never before.

In contrast, the UK has, since the 1990s, had a high degree of consensus about the need to address climate change. As I have described, in the late 1980s, Margaret Thatcher was one of the first national leaders to call for action. Since then, all major political parties have accepted the scientific consensus and made a case for action of some sort.

The big breakthrough in climate action came in 2008, with the introduction of the Climate Change Act, alongside the creation of independent adviser and watchdog body, the Committee on Climate Change. The Act was a product of political opportunity. Campaigning by environment groups, who wanted climate targets written into UK law, coincided with a particular set of circumstances. Ed Miliband, then Secretary of State for Energy and Climate Change, was a powerful and ambitious figure who understood the issue well, and wanted to make his mark. Meanwhile, David Cameron, leader of the Conservative Party which had been out of power for many years at that time, saw support for climate action as a way of showing the Conservatives as a caring, centrist party (Lockwood, 2013). As a result, there was cross-party support for the Climate Change Act, with only three out of 650 Members of Parliament voting against the legislation.

The Climate Change Act remains the most comprehensive piece of legislation that a national government has ever introduced. In June 2019, again with strong cross-party support, Parliament voted to amend the Act, implementing a target of the UK achieving net-zero emissions by 2050, at the recommendation of the Committee on Climate Change. There was unprecedented attention paid to climate in the 2019 general election campaign, too.

And yet there are serious shortcomings in the UK's approach, as I will discuss in Chapter Five. While government as a whole is committed to a zero-carbon target, it is not clear what responsibility different parts of government, or different sectors, have for achieving it. Overall emissions have declined in line with the targets, yet the Committee has criticized government for inadequate policies to meet future targets, particularly in buildings and transport, where emissions continue to rise. Because the parties have tended toward consensus on climate, there has not been a lot of discussion about the best ways to meet the targets. As I will discuss in Chapter Five, there is a tendency toward climate action by stealth.

What about China, the world's biggest carbon dioxide emitter? Its record is mixed. China is very reliant on coal-fired electricity generation. It plans to reduce the amount of coal-fired generation in the national energy mix, but is also providing finance for coal-fired generation elsewhere in the world. It leads the way in some green industries, as the world's largest manufacturer of solar panels, with a considerable stake in electric vehicle manufacture. Working with the US under President Obama, Chinese diplomacy was instrumental in securing the Paris Agreement in 2015, but the plan it has proposed to meet its Paris commitments is judged by the Climate Action Tracker to be 'highly insufficient' (Climate Action Tracker, 2019a).

China matters not only because of the sheer size of its economy and emissions, but also because it is seen by some as an alternative model for climate policy – so-called 'eco-authoritarianism' (Shahar, 2015). The view is that China's leaders, if they choose, can do what the science requires, and impose climate policies as they see fit, unencumbered by the need to win elections or gain popular support for policies.

The reality is somewhat different. Although China's leaders do not face elections, they do need to be seen as legitimate and competent rulers. The implicit deal is that the state will provide citizens with political stability and economic opportunities, and in return, citizens will accept the regime, not protest against it. This is sometimes called 'performance legitimacy' (Zhu, 2011). With the growing problem of air pollution and smog in Chinese cities, it may well be that environmental management is becoming part of 'performance legitimacy'. The government, in other words, may need to ensure environmental protection alongside economic opportunities, in order to fulfil its implicit pact with citizens.

So even though China is not a representative democracy, its ability to respond effectively to climate change is not as much of a contrast to democratic regimes as you might think. The

Chinese state may not rely on votes, but it does rely on a more implicit consent from its citizenry. Its ability to act on climate depends on whether citizens accept climate action as part of a wider mandate (Tyfield, 2018).

Next, Ethiopia. Around the time of the Paris negotiations, when I was working with climate groups to make the case for a strong deal, one of the most encouraging conversations I had was with a representative of one of the international development groups, who worked in London but was from Ethiopia. He described his country's climate strategy as 'leapfrogging'. Ethiopia wants to improve living standards, but acknowledges that following a conventional growth path would result in unacceptable rises in carbon emissions. So their strategy purposefully seeks out a low-carbon development path, which they have named the Climate Resilient Green Economy Strategy. The country is seeking partnerships and funding to pursue this route, particularly through increasing renewable energy. As a result, the independent watchdog Climate Action Tracker rates Ethiopia as one of the few countries whose strategy is compatible with global warming below 2°C, though it is not Paris compatible (Climate Action Tracker, 2019b). Meanwhile, the world's largest democracy, India, has similarly been ranked as 2°C compatible, although it remains the second largest coal consumer, after China (Timperley, 2019).

A tour through national climate strategies would not be complete without reference to those countries whose very existence is threatened by climate change – the small island states of the Pacific, already suffering from the consequences of sea-level rise. These countries could disappear entirely if global warming is not brought under control. Just before the failed Copenhagen summit, the Prime Minister of the Maldives, Mohamed Nasheed, decided to hold a cabinet meeting underwater, to make this point in the most direct way possible (BBC News, 2009). For the Maldives and other small island states, international climate negotiations are inextricably intertwined with their future as a nation.

This quick survey shows how differently countries have responded to the global climate negotiations. Two key lessons emerge. The first is that the different countries have different strategies, demonstrating that climate change is, most definitely, political. The second lesson is simple: no country has yet done enough. In the next chapter, I come on to the reasons for this.

Democracy in crisis

The core argument of this book is that climate action requires a social contract between government and people. It is a democratic challenge. I have not yet confronted an obvious truth: that democracy is in crisis.

The change in the atmosphere of politics and governance from when I started out working in and around politics in the late 1990s to the current situation in late 2019 is profound. For many years I, and others in my field, had been able to work with the assumption that the political system, though flawed in many ways, was based on certain standards of evidence and truth, and a sense of public duty. It would be an understatement to say that events since 2016 have thrown that assumption out of the window. The UK's decision to leave the European Union (EU) and the subsequent fraught state of UK politics, the election of President Trump in the US, President Bolsonaro in Brazil and the rise of nationalist parties elsewhere, have shaken the political establishment to the core.

When I began my research with politicians, in 2014, there were few overt signs of the political upheaval that was to come, though the seeds may well have already been sown, not least in the 2008 financial crisis and the responses to it. My first round of interviews was set against the backdrop of the EU referendum campaign, in late 2015 and early 2016, and I could sense an increasing uncertainty among my interviewees as the vote approached.

In the interviews I conducted following the UK's decision to leave the European Union, in 2017 and 2018, I noticed a growing diffidence among my interviewees. The Leave vote, Jeremy Corbyn's surprise ascent to leadership of the Labour Party, and the transatlantic ripples from Donald Trump's election, had profoundly unsettled UK politicians. The sense of disorientation and uncertainty was palpable, particularly among those from centrist traditions on both the Left and the Right.

Many thousands of words have been written about this new political reality, but here I will restrict myself to the question of what it means for climate action. There are two, linked, characteristics of recent politics which have profound consequences: first, the question of trust in government; and second, the status of expertise and evidence.

The first characteristic, the widely documented decline in trust of government (Foster and Frieden, 2017), has profound and deeply negative consequences for climate politics. In the UK, a hostility toward government manifests itself in the rhetoric of deregulation, cutting back 'red tape' and freeing up entrepreneurs and citizens to pursue their own agendas, unencumbered by the deadening hand of the state. This opposition to state intervention has been a key ingredient in Conservative politicians' criticism of the European Union, and in President Trump's pushback against environmental legislation.

Yet climate action, by its very nature, requires government intervention. This is partly because it is a problem of collective action – everyone benefits individually from the electricity, heat, transport or food that causes greenhouse gas emissions, but everyone suffers collectively from changes to the climate as a result of those emissions. A collective action problem requires a collective solution – at its very heart, it requires people to join together, in some form of government, to agree to take decisions in the common interest. But it is also partly because climate action requires changes to infrastructure (such as transport networks and electricity networks) that require funding or incentives from the state. In short, climate is an interventionist agenda, and this is at odds with the 'small government' mentality of current politics.

Second, a defining characteristic of recent politics has been the rejection of established expertise, and a demotion of the role of evidence. This sometimes strays into the promotion of outright falsehoods, as has been well documented in the statements of President Trump (Dale, 2019), and in the Vote Leave referendum campaign (Reuben and Barnes, 2016). Yet climate change is only fully understandable through scientific understanding and evidence. Climate systems at a planetary level can only be viewed through techniques of scientific observation, synthesis, modelling and forecasting, over long timescales and wide geographies; and the effects of emissions of greenhouse gases can only be understood with the help of these scientific techniques (Jasanoff, 2010). Climate action, meanwhile, measures its success through carbon accounting, and aggregated accounts of emissions reduction, at several steps removed from the action itself.

Thus an understanding of climate relies on the acceptance of expertise and evidence – which some people, and some politicians, are turning their backs on. This is one area where there is a clear political divide: it is right-wing politicians, and their voters, who are far less likely to accept the scientific consensus on climate change. Recent research for the Pew Center showed that less than a third of Republican voters agreed that 'climate scientists can be trusted a lot to give full and accurate information on causes of climate change', and among conservative Republicans this fell as low as 15 per cent (Funk and Kennedy, 2016).

All this leads me to believe that climate action is extremely vulnerable to the wider sweep of anti-expert sentiment, and the destabilization of established centres of knowledge and power. It should not, therefore, come as a surprise that there are clear links between these new political impulses and opposition to climate action. Investigative journalist Kyla Mandel has mapped the extensive links between Brexit-supporting lobby groups and the think tanks who question the scientific consensus on climate change, and notes that these groups even share an office building in Westminster (Mandel, 2016).

A further trend that populist politicians have been able to exploit is the sentiment that politicians and other 'elites' no longer understand or listen to people. There is some truth in this. Social researcher Claire Ainsley, in analysing responses to the UK's Brexit vote, argues that many politicians and policy experts have overlooked people's concerns and interests, instead trying to impose an agenda (Ainsley, 2018). My experience of the climate community is that they do the same. When I began working on climate change issues, in the late 1990s, climate advocates were small in number, and the debate was dominated by the more radical civil society organizations. Today, there is a well-developed climate 'establishment', which consists of many thousands of individuals working in finance, industry and government as well as civil society. Climate 'experts' are everywhere. One example is the renewable energy industry, now employing 10.3 million worldwide (IRENA, 2018).

This is a step forward, in that these thousands of people are focused on climate change (though their influence is still small compared to the enormity of the issue). Yet as this group of experts has grown, I have noticed an increased tendency to conduct discussions in the technical realm, without reflecting back on the need to develop a democratic mandate. There is a sentiment of doing this 'to' people, who, it is assumed, will either not notice or will passively accept change. Again, this makes progress on climate particularly vulnerable to political turmoil. In subsequent chapters, I will discuss a new approach to climate policy: one that starts from the expectations and understandings of citizens, not one that is imposed by experts, however well meaning.

THREE

The Energy Elephant

The scientist Jeffrey Dukes was driving through the deserts of Utah on his way to a research station a few years ago. As his car ate up the miles, he began thinking about the fuel in the tank, and the plants that it had come from. How many ancient plants, he wondered, had it taken to power him across the desert? He asked around, but couldn't find out. "The more I searched, the more frustrated I got. No one knew the answer" (quoted in Willis, 2011). So he did the sums himself. He worked out that a staggering 25 tonnes of plant matter go into every single litre of petrol. Imagine that: if you drive an average car, you would need 25 tonnes of plants every ten miles or so. That's six times the weight of your car.

The fossil fuels that power our cars, and that still provide 70 per cent of the energy we use worldwide (International Energy Agency, 2018a), are the remains of plants that grew millions of years ago. Those plants used photosynthesis to turn sunlight into carbon; over millions of years, this was condensed into coal, oil or gas, below the earth's surface. We started to dig it up in any serious quantities around 500 years ago – and we've been using it at quite a rate. According to Dukes' sums, the amount of fossil fuel we use per day is roughly equal to all the plant matter that grows on land and in the oceans over a whole year.

Dukes' maths had a profound effect on him. "I realized," he told me, "that nearly everything I do depends upon plants that grew millions of years ago; and that without them, my life would be completely different." The modern world that we take for granted is shaped by what Dukes memorably calls "buried sunshine" (quoted in Willis, 2011).

As Dukes shows, contemporary life in industrialized countries depends on a constant flow of fossil energy. In the UK, we each use, on average, fifteen times more energy than we did before the Industrial Revolution. My favourite illustration of our energy use is the most eccentric of all. The average US citizen uses so much energy that, if it were food, they would each be eating as much as a 27-metre-long brontosaurus every day (Catton, 2011).

In short, we use a huge amount of energy. But we barely notice. A plentiful supply is so ingrained in our lives and societies that we find it hard to see. That's why this chapter is called the 'energy elephant'. This issue's in the room, but we don't talk about it. In fact, it's been in the room with us for so long that we barely notice it. But it wasn't always there.

This only becomes obvious if we compare across time. As recently as the 1950s, the daily average distance travelled was five miles. Today, we each travel an average of 18 miles a day, not including flights (Department for Transport, 2019). My father, the son of a vicar, grew up in a village in the Yorkshire Dales in the 1940s. There were two cars in the village, belonging to the doctor and a rich landowner. My father's family didn't need a car, because it wasn't normal to own one. Work, school and shops were local; there were buses to the nearest town. Holidays meant a train trip to an aunt on the Yorkshire coast.

Today, living in that same village without a car would be a struggle. It would be difficult to work, shop or socialize without one; the expectation is that people have cars, and life is designed around them. A middle-class family like my Dad's would also expect to fly to a holiday abroad.

I tell this story not to say that everything was better in the olden days, but to illustrate the fact that the amount of energy we use is, to a large part, determined by the world around us. We can make changes, of course, but our own actions as individuals are influenced and constrained. The sociologist John Urry (2011) used the term 'socio-material systems' to describe the blend of physical infrastructure and social practices which provide the setting to our lives. The 'mobility system' of my father's boyhood in that Yorkshire village – a walk to the village school, shopping locally, and a holiday on the nearby coast – was very different to the mobility system of that same place, seventy years on.

In short, the amount of energy used by humans has been increasing steadily, year on year, for thousands of years (Berners-Lee and Clark, 2013), influencing our economies, societies, cultures and everyday lives.

It is important to acknowledge the huge benefits that modern energy systems bring to our lives. Nearly a billion people still do not have access to electricity, and this has severe impacts on their quality of life and on their ability to work and study (International Energy Agency, 2018b). Yet this relationship is not linear. Beyond a certain level, increased energy use does not improve quality of life. The average US citizen uses double the amount of energy as an average European (World Bank, 2014). US citizens are not twice as rich, and certainly not twice as happy, as Europeans. They use double the amount of energy because they drive further, and have larger houses with higher power needs (Willis and Eyre, 2011).

Technology to the rescue?

You might, at this point, say that it is not problematic to be using so much energy, as long as it is generated without carbon, from renewable sources like wind or solar power or nuclear energy. It doesn't matter how far we drive or fly if we use electric vehicles or planes, powered by renewables. 'Electrify everything!' might be the rallying cry. There are other potential solutions, like biofuels for planes, and a range of technologies proposed but not yet implemented, which could remove carbon dioxide from the atmosphere.

Certainly, decarbonization of industry, products, transport and so on is both necessary and possible. Biofuels could be an option for some types of transport. But there are two problems with the 'electrify everything' approach. First, it requires a shift of epic proportions, from fossil fuels to carbon-free sources of energy, which in turn requires huge investment and uses land required for other things like food production, tree cover (itself essential to a climate response) and habitats for other species (Berners-Lee, 2019). Trying to switch to carbon-free energy supply without addressing both energy demand and the drivers of that demand is a huge waste of resources, and stretches the limits of feasibility (Capellán-Pérez et al, 2019). Neither does it address questions of fairness, between countries or within countries. Simply switching technologies is unlikely to solve the deep inequity in access to energy.

The second problem is more subtle. As John Urry explains in his concept of 'socio-material systems', high-carbon systems should be understood not just as technologies and physical infrastructure, but also as social and cultural systems, influencing our expectations, practices and ways of thinking – including, of course, our politics (Urry, 2011). And because fossil energy is so ubiquitous in our physical world, it exerts a strong influence over society, culture and politics.

Political theorist Timothy Mitchell describes how the huge increase in oil use has resulted in a particular type of politics, which he calls 'carbon democracy', closely linked to economic theories that assume an abundance of energy. Previous generations of economists were preoccupied with resource scarcity. Yet in the twentieth century, they began to assume infinite access to resources. As Mitchell writes, energy abundance 'allowed economists to abandon earlier concerns with the exhaustion of natural resources and represent material life instead as a system of monetary circulation – a circulation that could expand indefinitely without any problem of physical limits' (2011, p 234). This economic outlook in turn led to a form of politics which did not consider energy or environment as a limiting factor: politics became, in Mitchell's words, 'dematerialised and de-natured' (2011, p 235).

Both Urry's and Mitchell's analyses show that the dependence of rich industrialized countries on a ready flow of energy is not just a technical issue, to be solved by technological substitution. It is, instead, something which conditions the way we think about our economy, our society and our politics.

Power politics and vested interests

Nearly fifty years ago, a political scientist called Matthew Crenson reported from the small town of Gary, Indiana, home to a huge steelworks plant (Crenson, 1971). The town of Gary had been established to provide a home for workers at the plant, by the company US Steel. The steelworks and the town had grown up together. In the 1960s, as concern grew about the health effects of air pollution, politicians across the US came under increasing pressure to introduce laws to clean up the air. Except in Gary, Indiana.

In nearby towns, Crenson documented the discussions, arguments, votes and laws that local politicians developed to deal with air pollution. Some politicians wanted strong controls on the factories that were polluting the air; others opposed these, arguing that these controls would make the factories too expensive to run. In many towns, local people got involved too. They argued for or against the new measures, talking about the health problems they faced, or the jobs that they feared might be lost if laws were introduced.

In the town of Gary, none of this happened. There was no debate. Instead, there was silence. The town carried on as if the air pollution wasn't happening.

Why this difference? Crenson's conclusion was fascinating. Of all the towns suffering from air pollution, Gary was the one most dominated by the steel company. Everyone there had family members employed in the steelworks; the politicians all had connections to it. Without the steelworks, there were no jobs. So no one mentioned pollution, because to address it would threaten the very existence of the town. That's why Crenson called his book *The Un-Politics of Air Pollution*. He wanted to make the point that, in politics, we should listen to the silence as well as the noise.

Crenson's study is about power. He shows that power works in slippery ways. It conditions how we think. So far in this chapter, I have described how carbon-intensive our lives are, in rich countries at least, and how this conditions not just the way we live, but the way we think. Crenson wrote his book to prove a point, that power should not be thought of just as formal control through laws and institutions. Power also manifests itself in what is said, what is not said, and what is assumed. Immersed in a high-carbon society, it is very difficult to think differently. When it comes to climate and greenhouse gas emissions, we are all citizens of Gary.

I have already shown how we are embedded in a high carbon society, which influences how we think and act. This is, to an extent, inevitable. But it is exacerbated by very deliberate strategies from those who have a stake in high carbon activities: countries, and companies, who depend on fossil fuels. Fossil fuel interests have three distinct strategies. The first is to question the science; the second is to position themselves as part of the solution; the third is to influence political and cultural norms. Fossil fuel companies do all three at once.

The historian Naomi Oreskes has charted in painstaking detail how corporate interests have employed a deliberate strategy of questioning scientific evidence, in order to create enough uncertainty to avoid legislation. These companies are, in the words of her book, *Merchants of Doubt* (Oreskes and Conway, 2012). This is what tobacco companies did, and it is what fossil fuel companies are doing now. It is well documented that Exxon had a good understanding of the science of climate change. In 1982, a memo was sent to senior managers, setting out the evidence base for climate change, and discussing 'potentially catastrophic events' such as the melting of the

Antarctic ice sheet (Exxon, 1982). This melting is, indeed, now occurring (Shepherd et al, 2019).

Yet Exxon then spent many years, and many millions of dollars, funding other organizations to question and deny the very same climate science. In 2015, investigations revealed that Exxon had directly funded a long list of private think tanks and policy institutes which question the scientific consensus on climate (Nuccitelli, 2015). The practice continues to this day.

While questioning the science, or paying others to question it, fossil fuel companies also position themselves as part of the solution. I have had many promoted posts on my Twitter timeline informing me that Exxon is investing in research into biofuel from algae. Exxon claims that, if the research is successful, the method could produce 10,000 barrels of algal biofuel per day (Peters, 2018). There are reasons to doubt these promises (Teirstein, 2018). But even if they are taken at face value, they need to be compared with the 2.4 million barrels of oil (or equivalent fuel) that Exxon produces every day (Garside, 2019). Exxon may view its biofuels operation as a small price to pay for a cultural licence to operate.

I might sound cynical at this point. But it is important to look at the past record of fossil fuel companies. Nearly twenty years ago, when I started working in this field, BP caused a stir by branding itself as 'beyond petroleum' (Macalister, 2000), investing in solar energy, as well as a new sunflower logo. At the time, I remember a genuine sense of excitement that such an important company could be setting out on a pathway of transformation. Yet its investments in solar and other renewable technologies continued to dwarf its core business, and in 2009, it quietly closed its renewables arm, BP Alternative Energy. Recently, however, it has had another go, reinvesting in solar and acquiring companies overseeing the rollout of electric vehicles. It now claims that it wants to make its operations consistent with the Paris Agreement – yet it continues to drill. How is this compatible with global carbon targets? As the ever-astute green business commentator James Murray points out, BP's 'Paris-compatible' strategy conveniently assumes that it is BP's remaining oil reserves, not anyone else's, that will be burned. Of course, if the other oil companies make the same assumption – that their oil and gas is the only stuff being extracted – then the Paris goal is unachievable (Murray, 2019).

The bottom line is that the share price of BP, Exxon and other oil and gas majors is based on a valuation of its reserves. If they cannot exploit those reserves, their value collapses. That collapse would send shockwaves through financial markets. The think tank Carbon Tracker coined the phrase 'carbon bubble' to describe the fact that financial markets are based on the assumption that known fossil fuel reserves will be burned – even though countries pledged through the Paris Agreement that this will not happen (Carbon Tracker, 2011). It's not just these companies that stand to lose if the oil stays in the ground – the economic model of whole countries, like oil-producing giants Saudi Arabia, Venezuela and many others, will be broken.

Not surprising, then, that BP claims its core business is, somehow, compatible with the Paris Agreement. Not surprising that the negotiators that oil-producing states send to international climate meetings actually work for oil companies (Climate Tracker, 2018). Not surprising that oil companies spend millions sponsoring educational and cultural institutions, to eke out their licence to operate that little bit longer.

Fossil fuel influences are pervasive across our culture and economy, but also have a direct influence on politics. Australian politician Scott Morrison, now Prime Minister, showed just how direct when, in 2017, he brought a lump of coal into Parliament to protest against his opponents' support for renewable energy. His administration, like the Republicans in the US, is heavily supported by fossil interests.

This isn't just a problem for right-wing parties. On the left, many politicians are reluctant to provoke trades unions who represent fossil fuel workers – hence the conflicting support for fossil fuels alongside renewable energy that is seen in Scotland, for example (discussed further in Chapter Four). There are countries like Saudi Arabia, where delegates to climate negotiations work for the state oil company (Hickman, 2018). And there are also more subtle influences: Norway has done more than any other country to invest in electric vehicles and renewable energy – and yet the investment comes from selling oil and gas. (To its credit, Norway recently announced that it would divest its sovereign wealth fund from fossil fuels [Kottasová, 2019].)

As these examples show, current political systems are a product of fossil-dependent economies and societies. That doesn't mean that change is not possible. A cornerstone of US Congresswoman Alexandria Ocasio-Cortez's election campaign was a refusal to take money from fossil fuel companies, or other large corporate donors. But it does mean that the complex intertwining of politics and the fossil economy needs to be acknowledged and challenged.

Will climate crises become political crises?

So far, I have described the ways in which politics has been shaped by the extraction and use of fossil fuels. There is another dimension to be addressed: as the climate starts to change significantly, what will this do to politics? Will climate crises become political crises? The earth is already locked into a period of warming as a result of greenhouse gases already in the atmosphere, however much we manage to reduce future emissions. The consequences of climate change are real and growing. Philip Alston, the United Nations' Special Rapporteur on extreme poverty and human rights, recently released a report which predicts huge rises in poverty as a result of climate impacts. As Alston writes:

> The uncertainty and insecurity in which many populations will be living, combined with large-scale

movements of people both internally and across borders, will pose immense and unprecedented challenges to governance. The risk of community discontent, of growing inequality, and of even greater levels of deprivation among some groups, will likely stimulate nationalist, xenophobic, racist and other responses. (Alston, 2019)

This is consistent with findings from the security community, who consistently identify climate change as a significant security risk or threat multiplier. The UK Ministry of Defence, for example, references the likely significant impact of climate change on patterns of migration (Ministry of Defence, 2014).

It is difficult to predict exactly how climate impacts will play out politically, either within countries or between countries. But it is safe to say that these dilemmas will become more central to political life over the years ahead. This is not something that politicians discuss much. Agonizing over uncertain futures, and speculating about geopolitical instability, is difficult territory for politicians. In my interviews, while some politicians talked about direct impacts, like sea-level rise or extreme weather events, they didn't speculate about how these cumulative crises could affect politics. Yet it is obvious that they will.

The political scientist William Ophuls has, for many years, charted the relationship between political arrangements and the natural resources available to human societies. He writes that modern political and economic thinking began with the advent of what he calls the 'age of abundance', beginning with Europeans' 'discovery' of the New World 450 years ago. European expansion into the rest of the world, through exploration, colonialization and, later, economic power, provided abundant energy and resources.

Ophuls argues that this abundance created the conditions for social and economic liberalism. Liberal democracy is predicated on choice – the idea that people, and states, have multiple options open to them, and so political decision-making is about deciding which options to pursue, in order to allow citizens, in turn, to make their own choices (Ophuls, 1992; see Dobson, 2013 for a discussion). If scarcity and environmental instability, rather than reliability and abundance, become the defining features of our societies and economies, this has profound implications for our politics.

Echoing this conclusion that liberal democracy is predicated upon environmental stability, political theorists Joel Wainwright and Geoff Mann (2018) speculate that there are four possible scenarios for global politics in a climate-changed world. The first they call Climate Leviathan, in a nod to Thomas Hobbes' philosophy. In this world, the power of global capital elites is strengthened and, through these elites, there is global cooperation to prevent and manage climate crises – though in a way which entrenches the privilege of elites, and increases inequalities within and between countries. Interestingly, this description has similarities with the 'Spaceship Earth' metaphor that scientists are drawn to, as I discussed in Chapter Two, though Climate Leviathan is an altogether more sinister vision. In the Climate Mao scenario, by contrast, there is equally strong global cooperation in the face of climate change, but this is managed by strong states rather than capitalist elites: Wainwright and Mann describe it as 'rapid, revolutionary, state-led transformation' 2018, line 841) and speculate that it could come about through climate-based disruptions to economic growth, in China and elsewhere in Asia.

In the two further scenarios that Wainwright and Mann set out, there is no global cooperation. The scenario they call Climate Behemoth is based around a reactionary conservatism, in direct opposition to international norms, which they see the seeds of in the regimes of Donald Trump and Narendra Modi. Last is their preferred scenario: Climate X, a trajectory which they name as utopian, but purposefully do not define, beyond saying that it is post-capitalist, and based around notions of equality, dignity and solidarity. It is more defined by what it is not, than what it is, and this, they say, is deliberate:

The planetary crisis is, among other things, a crisis of the imagination, a crisis of ideology, the result of an inability to conceive any alternative to walls, guns, and finance as tools to address the problems that loom on the horizon. (Wainwright and Mann, 2018, lines 3811–13)

Whether or not you agree with Wainwright and Mann's analysis or prescription, it opens up a much more explicit debate about the interactions between earth systems, human societies, and the politics that govern those societies. As this chapter has made clear, the way in which our material world is, and will be, shaped – by fossil fuels, by the retreat from them, and by climate impacts – in turn shapes the sort of politics we can expect or aim for.

Yet just as the material world shapes the political and social world, so the reverse is true. It sounds bleak to suggest, as I have done in this chapter, that the structure of the economy is intrinsically linked to our social worlds and outlooks, even our imaginations. Yet this acknowledgement contains its own escape route. If we face up to the energy elephant in the room, talk about it, and imagine a world beyond it, we can challenge these assumptions.

That is exactly what is starting to happen. Arts activists are taking a stand against fossil fuel companies' sponsorship of the arts, and scored big wins in the UK when the Royal Shakespeare Company and the National Theatre announced they would no longer take sponsorship from oil companies (Parveen, 2019). Extinction Rebellion's first demand, memorably painted on the side of a pink boat in London's Oxford Street during the protests in April 2019, is to 'tell the truth'. Acknowledging the energy elephant is half the battle.

FOUR

Dual Realities: Living with the Climate Crisis

When life gets me down, I go running. I have a collection of comedy podcasts which I plug into, as the dog and I make our way round the local hills at rather a sedate pace. My favourite is a show that is as old as me: the BBC's *I'm Sorry I Haven't A Clue*. It's a panel of very clever, very funny people doing very silly things. For me, it is the best medicine for climate anxiety.

To live in a time of climate crisis is to compartmentalize. If, like me, you spend many of your waking hours thinking about climate, it exerts a heavy toll. The news of what is already afoot, the wildfires, heatwaves, droughts and floods. The predictions for the future – within my own lifetime, and in the lifetime of my children. The intransigence of the response from politicians, media and many people. It goes round and round in my head, and I have to switch off. When I take time off work, I can feel myself disconnecting from climate change too, and it is a relief.

Responding to climate change is about balancing this dual reality: acknowledging the enormity of climate change, without being overwhelmed. But it is a difficult balance. Those of us who work on climate daily are stalked by it. Others keep it at a distance, or laugh it off with quips about the end of the world.

One of the frustrations of my research is that many university colleagues do not share my sense of urgency or immediacy. I vividly remember, early in my project, sharing some initial findings at a university seminar. I thought long and hard, and decided to talk to my colleagues about the dilemma that I discussed in Chapter Three: given our dependence on huge amounts of energy and on a stable climate system, how can we imagine or discuss a climate-changed future? My colleagues listened politely, but I found the discussion frustrating. They seemed to be locked into their safe, stable worlds, wanting a good intellectual debate, but not wanting to take this leap into imagining, or asking fundamental questions about our existence on this planet.

It seems that our politicians, too, are reluctant to open up these questions. My research showed a marked tendency to play down the climate threat. In this chapter, I look at why this might be. I start with the politicians themselves, and then look at wider public opinion.

Freaks and zealots? Climate change and the culture of politics

Julia (not her real name) is a confident politician and expresses her views freely. As we chat over coffee, she is deliciously unguarded in her opinions of her colleagues, criticizing the vast majority of her fellow parliamentarians for not dedicating time or attention to climate matters. She says that only a few of her six hundred or so colleagues take the issue seriously – "you might not get into double figures". For Julia, taking climate change seriously:

> '… means almost like seeing everything through that prism. So for example when the budget announced more tax breaks for North Sea oil and gas exploration, my immediate reaction is, but we're meant to be keeping fossil fuels in the ground to meet our carbon targets.'

She's right, of course. The science backs up her view that nearly all of the fossil fuels below ground need to stay there, if the world is to avoid dangerous warming. And yet Julia knows that she must tread carefully, not for scientific reasons, but sociological ones: she has to fit in. I ask her what would happen if she tried to persuade her colleagues that the fossil fuels should stay in the ground. She replies: "I think they'd think you were a bit 'niche', is the way I'd put it – I say 'niche' in quotes like a bit of a lunatic fringe."

Julia isn't the only one who worries about her 'niche' reputation. One former MP, who had been an active climate campaigner in Parliament, said "I was known as being a freak". Another told me about how he tried to avoid being seen as a "zealot". He said he had been arguing for better public transport in his constituency, and I asked him whether he had mentioned climate change. He said he hadn't: "I think if I had mentioned carbon emissions, there would have been a rolling of eyes and saying, 'oh here he goes again'." These remarks are common in my conversations with politicians. Some went as far as deliberately avoiding any mention of the climate, for fear that it would be an unhelpful label.

This shouldn't come as a surprise. As any undergraduate sociologist learns, the way people think and act is conditioned by their social world (Lawler, 2014). That's not to say that people have no free will. Of course, we can challenge, question and think for ourselves. But we are heavily influenced by our social surroundings, and by implicit rules and norms. If you start a new job, you need to find out pretty quickly what your colleagues talk about in breaks or how meetings work. In my experience, in some organizations, meetings always start on time. In others, it's perfectly fine to turn up five minutes late and chat before starting ten minutes after the scheduled time. The first few days in a new job are difficult, because you don't know the rules. Social rules are rarely explicit, but that doesn't lessen their power.

Like any workplace, the UK's House of Commons, where I conducted this research, has strong social norms and implicit rules. In the 1990s, Nirmal Puwar (2004) spent many hours talking to female politicians and those from ethnic minority backgrounds about how their race and gender influenced their work in the Commons. They reported strong pressures to conform. They felt like outsiders, because of their gender or ethnicity, and were faced with a choice. They could either accept the rules of the game, and try to be 'like any other politician', or they could choose to stand out, talk about issues affecting women or ethnic minorities (either within Parliament or more widely in society), and be seen as an outsider. The fascinating point, though, was that white male politicians didn't feel these pressures, and didn't notice the hidden rules that governed their natural habitat. They just assumed that their way of being a politician was the natural, or neutral, way of going about things.

When I read Caroline Lucas' (2015) account of entering the House of Commons as the only representative of the Green Party – a position she still occupies at the time of writing – I was struck by her description of the isolation she felt. It wasn't just political isolation; she would have expected that. It was personal and social too. She just didn't fit in. Colleagues were often patronizing and dismissive. When she talked to a fellow politician about her ideas for reforming the MPs' voting system, the response she got was "you just don't understand how things are done here".

So what does this mean for climate change? My research with MPs demonstrated clearly that taking an active role on climate does not fit current institutional norms. Imagine that you are a politician who understands the need for urgent action on climate. You have a choice of three strategies.

First, you could speak out at every opportunity – at every debate about the economy, or transport, or housing, you could ask how this will help to meet carbon targets. You could be very critical of colleagues and proposals that don't take climate change seriously. You could put forward suggestions for laws or policies that would tackle the issue – for example, a ban on new fossil fuel extraction. Doing this has the benefit of simplicity and directness – saying it like it is. But it has a big downside: you will be regarded as an outsider, a freak or a zealot (these words were all used by my interviewees). You will be dismissed as a radical, as someone not in touch with reality.

As the sole MP representing the Green Party, Caroline Lucas plays this role, of course. That's partly because she doesn't really have a choice. She's obviously an outsider. She asks awkward questions whenever she can, and her aim is to unsettle the consensus. But for members of the bigger parties, it's different. Politicians told me that this strategy affected prospects for promotion. If you are too forthright, you may not be seen as a suitable candidate for ministerial office.

When I talked to people from environmental groups who work with politicians every day, they were very frank: they said that climate change wasn't an issue that would help politicians establish their reputation as a serious player. One environmentalist described how it might look to a government minister wanting to develop their personal power and influence:

'If you're [a minister] and you broadly think that climate change is happening and you should do something about it, you walk into Cabinet and you start saying "right guys, what are we going to do about climate change?", you'll just get laughed out of the room. They want to be talking about the economy, and building stuff, and bombing people. It's just not a serious sort of Cabinet issue for the big bruisers. If you're trying to build your base in a party, you absolutely don't do that by talking about airy-fairy climate change. You do that by talking about jobs and the economy.'

So if you don't want to be too direct, what other strategies are there? The second option is to see what you can manage to do on climate change, but just tread more carefully, and try not to be too radical or confrontational – try to fit in. Quite a few of the politicians I spoke to did this, including Julia, who I've already described. She really wanted to make a difference, but she judged that being too outspoken could backfire. Like many others, she tried to find ways to broach the subject, maybe focusing on the economic benefits of renewable energy, while also mentioning the climate benefits. She would support colleagues who suggested small steps forward, instead of criticizing them for being too cautious. She would try to build alliances, and take people with her, even if that meant diluting her own message a bit.

When I analyzed the words that politicians use when speaking about climate in public debates, a key finding was that there is a strong tendency to dilute the message. Politicians were reluctant to speak out about the severity of the climate problem, or the far-reaching changes to our society and economy that will be required. In nearly a hundred thousand words of debate about climate change in 2009, there were only three mentions of abrupt or irreversible impacts, often called 'tipping points', even though these are widely discussed in the scientific literature. The politicians appeared to be presenting climate change as a relatively unthreatening, manageable problem. By defining it in this way, it made it easier to talk about. It is a tactic, a well-meaning attempt to frame a difficult, complex issue into something amenable to the political agenda (Willis, 2017).

The last option, if course, is not to talk about climate change at all. You could either ignore it, channelling your energies into other issues; or try to support the agenda without shouting about it – the strategy that I call 'stealth climate action', discussed further in Chapter Five. Many politicians, though, might accept the need for action, but choose not to work on it, leaving it to other people. As one said, it is easy to ignore, as "it's just not part of the daily discourse, it's too far away". Another described it as something that "probably falls into the basket of general progressive issues that sound good to ensure". This isn't hostility or opposition, just a choice not to prioritize the issue. It's a choice which, until now at least, has been easy to make.

As climate impacts begin to bite, and protests grow louder, this situation might be about to change. Researcher Rhian Ebrey (2019) has analyzed the Twitter activity of UK MPs since September 2018. There has been a sharp rise in mentions of climate from the Twitter accounts of 577 MPs, from an average of 1 per day at the start of the period studied, up to 7.8 mentions per day in May 2019, peaking with 160 tweets on 1 May, when the UK Parliament declared a climate emergency. Ebrey cites the school strikes and young campaigner Greta Thunberg as the prompt for this upsurge. The research also revealed a change in language used, with a greater tendency for politicians on the left to use language of 'crisis' and 'emergency'. This seems to represent a significant increase in attention to climate change. Although it is too early to say whether the change will last, it may be that the culture of climate silence is now being overcome.

In the US, there has been a corresponding rise of climate change up the political agenda, driven in particular by the advocacy of Congresswoman Alexandria Ocasio-Cortez, and lobby group the Sunrise Movement. Again, change has been swift, with nearly all Democrat presidential candidates stressing their commitment to climate action, something never previously seen in the race for nomination. In the US, however, as I will go on to discuss, the issue is more partisan, with support for climate action strongly linked to the Democratic party, and little or no discussion among Republicans.

What about the voters?

So far, I have only talked about the influences that politicians feel within their own, peculiar workplace. But politicians are elected representatives. They don't just care about what their colleagues think – it is their job to get people to vote for them, and, once elected, to represent those people in Parliament. So the perceptions politicians have about the views and outlooks of the wider public, and particularly voters, will help to shape a politician's actions. That's what democracy is about.

The most common response to climate change is neither denial nor activism, but what public opinion expert Leo Barasi (2017) calls 'climate apathy'. There is a large group, the majority of UK citizens for example, who do not dispute the science, but don't do much about it. They don't talk about it much, they don't try to reduce their own emissions much, and they don't bear it in mind when voting. In early 2019, I helped Green Alliance to commission some 'citizens' juries' on climate change. In Cardiff and in Penrith (in North West England), we spent a day talking with citizens and their MP to get an idea how they felt about climate change, and what they wanted their government to do. Overall, participants were aware of and worried about climate, but they had no real sense of urgency or agreement about what's at stake (Buck, 2019). One person said that she saw "apathy and resistance to change. I think people are just happy with their lives and they don't want to get involved".

Why might this be? A significant factor is the social norms that I described above, which constrain politicians' action on climate. They apply to the rest of us, too. The researcher Kari Norgaard spent a year living in a rural Norwegian village, observing how the community talked about climate change. That winter, the weather was unusually warm. The lake didn't freeze over and the local ski area had seen no snow by mid-December and had to resort to artificial snow – something that had not happened before. Income from the ski industry was reduced, and the tradition of ice fishing was interrupted. When talking about this and other weather patterns, villagers mentioned climate change, and seemed to have a good level of understanding. And yet it was not something that people talked about or factored into their daily lives. Norgaard uncovered an interesting paradox. As she writes:

> From my direct observations and the reports of community residents in interviews, people were aware of the causes of global warming, had access to information which they accepted as accurate, yet for a variety of reasons they chose to ignore it. This was a paradox. How could the possibility of climate change be deeply

disturbing and almost completely invisible – simul-
taneously unimaginable and common knowledge?
(Norgaard, 2006, p 350)

Norgaard had thought that she was studying people's knowledge of climate change. She realized that she was, in fact,
studying something else. A different question began to shape
her research. She started to ask how people managed to live
their lives knowing that climate change was happening, but
acting as if it was not: 'How did people manage to produce
an everyday reality in which this critically serious problem
remained invisible?' (Norgaard, 2006, p 364).

At the level of the psychology of the individual, it is well
documented that individuals develop strategies of denial to
avoid confronting difficult realities. Someone who is deep in
debt may compound their problems by trying to ignore their
money worries, hoping these will somehow disappear, and
keeping on spending. Norgaard's observation, though, was
that the denial was social – it was shared among the whole
community. It was as if they unconsciously conspired with
each other so as not to think through how climate change
was affecting them or whether they should be making any
changes to their lives. Norgaard points to a range of strategies
they employed to allow them to continue as if climate change
wasn't happening. Some said that Norway was a relatively
small country, with insignificant emissions compared with the
US. There was also a widespread sense that, as a rural community, they lived simply and 'with nature'. By telling each
other these and other stories, they created the fiction that all
could continue unchanged.

A study by Melissa Petersen (2018) of the inhabitants of the
Republic of the Marshall Islands revealed a similar picture.
These islands, in the Pacific Ocean, are just five metres above
sea level, and so are uniquely vulnerable to climate change.
Many Marshallese have already left, and more are planning
to leave, as they have a right to residency in the US under

a military treaty. However, when Petersen interviewed the islanders about their reasons for moving, they stressed the economic opportunities of life in the US, rather than the threat to the Islands, as the primary reason for leaving. While they know and acknowledge that the seas are rising, this remains in the background.

Even those of us who work on climate change on a daily basis have our denial strategies – like my comedy podcasts. In fact, it may be even more necessary to develop coping mechanisms, to preserve psychological wellbeing. Psychologists Lesley Head and Theresa Harada interviewed climate scientists, to find out how they coped with the often disturbing evidence that they uncover. They found that the scientists 'distance themselves from stress and anxiety by downplaying the painful or troubling emotions' (Head and Harada, 2017, p 40) that they experience in working on climate. This helps them to keep going. But, like those politicians who ignore the most troubling implications of climate change, there are side-effects to this strategy. In trying to downplay the worst of it, they insert a bias into their thinking. They 'systematically downplay worst-case scenarios and embody a kind of everyday denial favouring positive scenarios' (Head and Harada, 2017, p 40). They want to feel that it is going to be okay, even if they know their evidence suggests otherwise.

It turns out, then, that the denial and avoidance that Norgaard identified affects us all: citizens, politicians, scientists. As one MP said to me, "The majority of MPs recognize that climate change is manmade, it is happening and it is going to have catastrophic consequences, but it's so scary in some ways, maybe they don't want to think about it".

2019: a climate spring?

Recently, something has changed. I've found that lots more people are talking about climate, and wanting to know about my work – friends that I know through my children's school, people I get chatting to on trains. Climate impacts, media attention and protest movements may be changing the tendency toward denial or dismissal of climate change that I described previously. Opinion polling has been showing startling increases in people's reported levels of concern about climate change. An August 2019 poll by Ipsos Mori, following the hottest month ever recorded on earth, showed that 85 per cent of UK adults are now concerned about climate change. This was an increase of 18 per cent since 2014, and the highest figure since the poll began in 2005 (Cecil, 2019). The UK isn't an exception: in a Pew Center study of 26 countries in 2018, climate change was seen as the 'top international threat' in half of those countries. In all but three countries, more than half of those surveyed identified climate as a major threat – the exceptions being Israel, Nigeria and Russia. In this survey, concern had increased significantly since 2013 (Poushter and Huang, 2019).

There is undoubtedly an increase in awareness of, and concern about, climate change, and the impacts on politics are already being felt. Yet the relationship between people's stated concern, as expressed to pollsters, and their actions, behaviour and voting intentions, is not straightforward. As Norgaard's Norwegian village demonstrates clearly, concern about climate change can coexist with contradictory beliefs and actions. During discussions in the citizens' juries I mentioned earlier, there was no sense of a critical moment, or 'emergency', despite the protests. Participants were confused why, if climate change was so serious, life – and politics – was continuing unchanged. Like the politicians I interviewed, they were struggling to reconcile the science of climate change with the social cues that they were getting from those around them. As one participant in Cardiff said, "I'm quite shocked by it, because it's something thirty odd years ago we didn't consider, and I thought nothing would come from it. Now the little ones bring it up all the time, and it does make you think".

Cool dudes and catastrophists

So far, I have presented a generalized picture of public opinion. But there are some crucial differences between groups as well. As we saw in Chapter One, in the UK, Europe, the US and Australia, views on climate correlate with political outlook, with right-wing voters and parties less likely to be worried about climate or to support action, and left-leaning voters and parties showing higher levels of concern. The landmark 'cool dudes' study in 2011 by Aaron McCright and Riley Dunlap revealed that conservative white males contribute disproportionately to climate denial in the US. They argue that the reasons for this are sociological:

> … conservative white males have disproportionately occupied positions of power within our economic system. Given the expansive challenge that climate change poses to the industrial capitalist economic system, it should not be surprising that conservative white males' strong system-justifying attitudes would be triggered to deny climate change. (McCright and Dunlap, 2011, p 1171)

Other US studies have also shown differences in attitude linked to gender, religion and age, with men, older people and people with religious beliefs less likely to accept climate science and support climate action. A study by Gregory Lewis and colleagues found clear political divisions on climate beliefs, with conservative voters considerably less likely to call climate change a serious issue than liberal voters. The biggest partisan divides occur in Australia and the US. There are lesser, but still significant, divides in Canada, the UK and European countries (Lewis et al, 2019). However, these trends are far from universal, with little evidence of polarization in many countries. Lewis' study found the only consistent factor across 36 countries studied was that people who had 'a strong commitment to democratic principles' were more likely to be concerned about climate change.

There is another group whose views differ significantly from the mainstream. There is a relatively small, but growing, group who believe that the impacts of climate change will have a catastrophic effect on human societies in the near term. Roger Hallam, one of the founders of the protest group Extinction Rebellion, recently stated in a BBC interview that he expected six billion people to die from climate-related causes (Hallam, 2019). Hallam and others within Extinction Rebellion argue that the impacts of climate change will be so severe that, soon, human societies in anything like their present form will no longer be viable. In a similar vein, the Dark Mountain Project draws together a group of artists and activists who look beyond current societies to a future of 'uncivilization' (Kingsnorth et al, 2014). They see their work as imagining and starting to build alternative futures, beyond the collapse of our current societies.

Thus climate deniers and climate catastrophists are at opposite ends of the very wide spectrum of public concern on climate change. For me, two questions emerge from this. First, how does this spectrum of public concern match with scientific evidence and predictions? And second, what are the implications for politics, and for developing a workable political strategy on climate?

To take the science first, existing scientific evidence offers no support for climate deniers. This is abundantly clear. As I described in Chapter One, climate change is already having significant, lasting impacts, and these will worsen over time, particularly if efforts to halt emissions are limited. To put it simply, on the basis of scientific evidence, concern about climate change should be very high indeed, and certainly a great deal higher than the level of concern expressed by most people, most of the time.

The question of whether there is evidence to support the climate catastrophist view is more problematic. As I have discussed throughout this book, there is strong evidence that climate change will reduce crop yields, decrease the availability of fresh water, and increase extreme weather events. It is also clear that this will result in a huge amount of human misery and poverty. Yet the ways in which this will unfold, and the impact on human societies in aggregate, are complex and difficult to predict. A 2014 World Health Organization (WHO) study predicted 250,000 additional deaths each year from 2030, from factors attributable to climate change, including heat exposure, diarrhoea, malaria and malnutrition. However, this study assumed 'continued economic growth and health progress' (WHO, 2014). If climate instability leads to political instability and economic collapse, outcomes could be far worse.

Very little academic work has been done on the link between climate change and civilizational collapse, partly because the immense variables make it difficult to study – and maybe because of the sort of 'optimism bias' that Lesley Head's study of climate scientists uncovered (Head and Harada, 2017). So evidence is limited, but it is impossible to say that the catastrophists are definitely wrong.

But in the political world, the picture looks different. Although their position is not supported by science, climate deniers have a solid foothold in mainstream politics, particularly in the US Republican Party, and Australia's Liberal Party, and also have political support in the UK and elsewhere in Europe. Climate catastrophists, meanwhile, have no such foothold. This is perhaps because it is much easier for politicians to support the status quo than it is for them to argue that our civilization is not viable. Even politicians who hold the most ambitious positions on climate action, such as radical Democrats in the US and Green politicians in the UK and Europe, ground their visions in an assumption of the continuation of complex human societies.

Some may see this as unfounded optimism. Yet it is also an acknowledgement that, when it comes to human societies, prediction is never just prediction. It carries within it a set of assumptions, prescriptions and hopes. For politicians, the challenge is to acknowledge the enormity of the climate crisis, while not writing off the world as we know it. We have to start from here.

So far, most politicians in most political systems have failed in this endeavour. In Chapter Five, I look at what efforts have been made, and how they have fallen short.

FIVE

Twenty Years of Climate Action – but Still Emissions Rise

As we saw in Chapter Four, public concern and political attention on climate have been limited. In this chapter, I will look at what this has meant for climate action. How do existing efforts to tackle climate change measure up against what the science suggests is needed?

A few years ago, I was invited to speak at the awards ceremony for a leading UK university. They had run a competition to identify and reward ten innovators from across the university who were 'saving the earth'. There were some brilliant projects. The student union had developed a scheme to encourage final-year students to donate pots, pans and other household items to new arrivals, saving resources and a trip to Ikea. The university's engineering department had been at the forefront of research into new forms of solar power. But when I spoke at the ceremony, I made myself unpopular. I was very kind to the winners, but I also pointed out that they should also compile a list of the ten most significant ways in which the university was wrecking the earth – through investing in research into oil and gas extraction, for example, as most of our leading universities still do.

Feelgood fallacies and stealth strategies

This story shows the first fundamental problem with climate action to date. This is a problem that I have come to think of as the 'feelgood fallacy'. There has been an overwhelming focus on encouraging low-carbon solutions – like developing renewable energy, or offering grants for electric vehicles. These are valuable things to do. I have worked on many projects to promote renewable energy, for example, and I support them strongly.

But all this positive activity masks a deeper problem. Very little has been done to curb carbon-intensive activity, like new sites for fossil fuel extraction, increasing demand for aviation, and growing meat consumption. The politicians I have spoken to are nervous about addressing these issues. Environmental campaigners have often told me that they worry about arguing for changes to aviation or meat consumption, because they worry it might alienate people.

It won't be enough just to ramp up renewable energy. Although renewable energy has grown at an impressive rate in recent years, fossil fuel use is rising too. Study after study shows that meeting the goals of the Paris Agreement means phasing out the extraction and use of oil, coal and gas (Berners-Lee and Clark, 2013; McGlade and Ekins, 2015).

The science is clear: emissions will keep on rising unless the really tricky issues of fossil fuel extraction, aviation and agriculture are faced head on. But there is a danger that people and organizations channel their energy into the easier sell, encouraging the positives rather than fighting the negatives. This is the 'feelgood fallacy'.

The second fundamental problem is one that has cropped up throughout this book. This is the assumption that experts know best, that they can impose solutions without people noticing or caring, and that their expert views override public opinion. Remember the politician I interviewed, who tried to make the case for local transport improvements, without even mentioning that they would save carbon? And the scientists who want to impose legally binding 'planetary boundaries' but have very little to say about whether or how these should be decided democratically? I call these attempts to tackle climate change 'stealth strategies', because they attempt to do what they think is best, and hope that no one notices.

Both these problems link back to the issues raised in Chapter Three, about power and vested interests. It's easier to concentrate on innovative new technologies, rather than picking a fight with powerful fossil fuel interests – hence the feelgood fallacy. It's easier to suggest small, incremental changes that won't challenge dominant social views, than trying to engage people in challenging conversations about social futures – hence stealth strategies.

This was brought home to me in a conversation I had with one politician, who represented a local area where there were lots of jobs in a carbon-intensive industry (I can't be more specific than that, to preserve anonymity). She played an active role in climate policy and debate in Parliament. She was arguing for stronger climate action but, at the same time, supporting initiatives that would benefit her local industry. I asked her whether she talked openly about the need for radical climate action. Her reply was intriguing. She said:

> The challenge that I would get is one of hypocrisy. I support doing things that will benefit [the local carbon-intensive industry], while saying we should take action on climate change. Very few people have ever raised that with me, it's more of an internal 'how do I justify this to myself?' ... I kind of thought I might get a bit of pushback from constituents, but I've had absolutely zero.

To my mind, this answer encapsulates the problem with our partial attempts to reduce emissions to date. Like this politician, we've encouraged some good stuff, but we haven't yet had an honest conversation about the power, the vested interests, or the choices we have to make if we are to achieve significant reductions in emissions.

Collectively, governments now have three decades or more of climate policies to reflect on. There has been a huge amount of detailed research on strategies, policies and innovation to tackle climate change. I couldn't hope to do it justice here.

What I want to focus on is what governments, at both national level and local level, have already done, and whether or not it has succeeded. I will look at three areas: first, the intersection between economic policy and climate change; second, local strategies for climate action, including local carbon targets and city-level strategies; and third, energy policy.

Economic policy and climate change

Professor Paul Ekins probably knows more than anyone in the UK about how to use the tax system to reduce carbon emissions. For more than thirty years, he has patiently explained, to successive generations of civil servants, the rationale of green taxation. The proposition is simple. Using taxation to increase the price of carbon-intensive activities, such as driving large cars or using fossil fuels to generate electricity, makes these activities more expensive than their less polluting alternatives, and so discourages people from choosing them while increasing their financial incentive to buy and invest in alternatives. There's an added bonus, too – the government can raise money for public services through these taxes on 'bad' things, meaning it could potentially reduce taxes on 'good' things, like income tax or company profits. In other words, a shift from taxing 'good' things to taxing 'bad' things would be economically beneficial as well.

Sounds good in theory. But in practice, it's a tricky thing to do. Taxes are never popular, and politicians mess with them at their peril. I remember vividly the fuel tax protests of 2001, sparked by a planned increase in fuel duty, the so-called 'fuel duty escalator'. Lorries blocked the streets, petrol supplies ran low, and the government backed down. Eighteen years later, the planned increases have still not been implemented. Recent protests in France by the *gilets jaunes* were also sparked by a carbon tax, with catastrophic results for President Macron. In both these cases, it is quite possible to argue that the proposed taxes were badly designed and not introduced properly, as part of a wider package of support. But any politician, looking at this experience, would be apprehensive about similar measures.

There's a certain irony to green taxation. The whole point of any such tax is to change the behaviour of individuals and companies. If the tax is set too low, it doesn't do this. But if the tax is set too high, people notice – that's the whole point – and may object. So the very point at which the tax begins to be effective is the point at which it becomes politically problematic, and it is very difficult for politicians to hold their nerve.

That's why, for as long as I can remember, governments have investigated the potential for green taxation without actually implementing any policies that might make a difference. Paul Ekins once told me, with a wry smile, that every few years, the government would ask him to sit on a committee or taskforce to investigate how economic policy could be used for environmental outcomes. Every time, he would reply with enthusiasm, and get to work recycling the evidence he had produced for previous such inquiries, in the hopes that this time, something might be implemented. He remains hopeful.

There does indeed seem to be a pattern to this. A few years ago, I counted up all the official government reviews, councils and taskforces there had been, over a decade from 1999 to 2011 (Willis, 2011). I found ten – that's at least one per year, and there have been more since then. All the reviews involved independent experts from industry, the third sector and academia. They all came out with very similar messages. To paraphrase: government needs to show leadership, and set a clear framework for transition to a sustainable, low-carbon economy. Polluters should pay more, and non-polluting alternatives should pay less, through tax shifts and investment support. There would be clear economic benefits to this strategy.

Yet very few of their recommendations have seen the light of day. Governments have used these independent reviews as a way of buying time, and claiming that they are about to act, as soon as the evidence is in. For politicians, investigating alternatives is less risky than implementing them. It is the ultimate feelgood fallacy. Yet it is also self-defeating. The less that politicians speak out and make a case for change, the less support they will get. In subsequent chapters I will investigate how politicians and others can put forward a positive, radical agenda that engages and motivates people, rather than trying to hide behind taskforces and enquiries.

Local strategies for climate action

Just before the UK's landmark Climate Change Act was passed, I moved to the Lake District, a beautiful rural area in North West England made famous by William Wordsworth's poems and Beatrix Potter's stories for children. It was quite a change from my London life. I began to think much more about how local areas fitted into the climate picture. I wanted to see what I could do in my new neighbourhood. So I went to meet the Lake District National Park Authority, whose job it is to manage this picturesque little corner of the world. I talked to them about the UK's carbon budget, established as part of the Climate Change Act, and about what that might mean for local areas. They responded with enthusiasm, and together we came up with a plan: we would have our own mini carbon budget at a local level, to mirror the national efforts.

This is how it works. The Climate Change Act specifies that the UK's carbon emissions must decline steadily, and requires government to produce a plan every five years, to meet the targets set out in the Act. But the Act doesn't specify what contribution local areas should make. In the Lake District, we decided to take on that responsibility for ourselves. We set ourselves the challenge of measuring greenhouse gas emissions from the whole local area, and reducing them in line with the national targets (Lake District National Park Authority, 2018).

Eighteen million people a year visit the Lake District. So we were not surprised to see that the biggest impact, nearly half of the total footprint, came from visitors travelling to and from the area, by plane and car. Car travel within the Lake District, by visitors and residents, was also very significant. Other sources of emissions included food and drink, as well as heat and power for houses and businesses.

Armed with this information, we set about looking at potential carbon savings. We convened a group of people, each representing the organizations who, between them, look after the Lake District – businesses, local government, and organizations like the National Trust, who own land and have a lot of influence locally. We discussed projects that could help shift us toward a low-carbon Lake District, which would also make it a nicer place both to visit and live in.

We have had some real successes. In 2012, we won funding from the Department for Transport for an ambitious travel strategy, now known as GoLakes Travel, for visitors to the Lake District. Over three years, bus services were improved, boat buses introduced on the lakes, cycle paths built and visitors provided with information on alternatives to car use. Visitors were even told, explicitly, to 'drive less, see more'. Results were impressive. Over the life of the project, car travel by visitors reduced by 14 per cent, with increases in walking and cycling. An estimated 42,000 tonnes of carbon were saved through GoLakes Travel. Other successes included a huge increase in locally produced renewable energy, particularly from small-scale hydro powered schemes (making use of the legendary Lake District rain) and schemes to help businesses reduce their energy and resource use. My personal favourite was a pub landlord in the Langdale valley who announced proudly that he was switching to locally brewed beer, which has a much lower carbon footprint as it doesn't have to travel very far from the brewer to the beer drinker.

It is important to celebrate the successes – the beer, the bike paths, and the businesses saving money on their heating bills. But despite these, it hasn't been an easy ride. The feel-good fallacy is alive and well in the Lake District. There have been tensions running through the whole project. An obvious issue is the plane journeys that people take to get to the Lake District. The tourism strategy encourages overseas visitors, and yet flying is the single biggest contributor to the area's carbon emissions. Every now and then, someone from a local environmental group points out these contradictions. Invariably, they are listened to politely, and then things continue as they were. There is widespread support for a new passenger airport just north of the Lake District, outside Carlisle, but there has been no official attempt to measure or manage its likely impact on carbon emissions.

Another obvious flashpoint is car use. The GoLakes Travel project is an impressive attempt to encourage more sustainable forms of travel. It went further than most, in openly encouraging people to drive less. But some obvious things that could be done have not really been spoken about. Cars could be charged to enter the Lake District, with the money spent on travel alternatives, much like the green tax shift that Paul Ekins has been promoting for so long. There have been quiet discussions for years, but local politicians have never felt able to talk openly about such a scheme, despite severe congestion on the Lake District's busiest roads.

Perhaps the most outrageous example of the feelgood fallacy that I've come across in my local work was when I met the local economic development agency for Cumbria, the county within which the Lake District sits. They said to me that they wanted Cumbria to be a climate leader, and they wanted my advice. We talked about some good things that they could do, like promoting renewable energy and supporting businesses to reduce their energy use. Then they mentioned the new coal mine they were supporting in West Cumbria. Mustering the calmest voice I could, I pointed out that opening up new coal mines was totally at odds with their desire to lead the way on climate change. It was awkward, to say the least.

My home patch is what I know most about, but the discussions and dilemmas that we have encountered in the

Lake District will, I am sure, be repeated in every area. Local climate leaders are not helped by a distinct lack of support from national government for their efforts.

Nevertheless, , the latter half of 2019 has seen a reinvigoration of local climate action, with many cities and local areas declaring a 'climate emergency' and setting ambitious targets for carbon reduction. In the US, state- and city-level action provides a counterbalance to the Federal administration's dismantling of climate policy. I will discuss this more in Chapter Seven.

Energy policy without people

I have spent a lot of time working on energy policy, looking at how government can create an energy system that provides the services that households and businesses need, without carbon emissions. One thing constantly strikes me: there is very little discussion of people in energy policy; it is all about technology and economics. When people are discussed, it is as 'consumers', wanting low energy bills. A century of energy policy has focused almost entirely on the supply side – how energy is generated, to the exclusion of the demand side – how energy is used (Willis and Eyre, 2011). So far, efforts to decarbonize the electricity system have focused on substituting high-carbon generation technologies, like coal-fired power plants, with low-carbon generation technologies, like wind energy and converting power stations to burn biomass (wood and plant matter) instead of coal.

There are other ways of saving carbon – through changes to home energy use, transport choices or consumption patterns. These could be helped by the wave of innovation now sweeping the energy system, particularly small-scale renewables like solar power, battery storage, and developments in information technology which enable much more sophisticated control over demand and supply. These new technologies provide opportunities to put people at the centre of the energy system, engaging them as citizens, not just as consumers. We could reduce energy demand and carbon emissions, in ways that improve our quality of life.

Yet the politics of energy, in the UK at least, is stuck in a rut. All the focus is on the supply side. There has been no attempt to develop policies that encourage or incentivize people to take an active role in the energy system, and there are few incentives to change consumption patterns or reduce demand. For many years, the assumption driving energy policy has been the stealth strategy – that low-carbon can substitute for high-carbon, without anyone noticing or caring. This assumption blinds policymakers to the huge opportunities, enabled by new technologies, to engage and motivate people to do things differently.

As these examples show, so far, most attempts by governments to reduce carbon emissions have been too limited in scope, not tackling the crucial issue of reducing fossil fuel extraction and use – what I have called the 'feelgood fallacy'. Governments have not sought to engage or motivate people, instead trying to decarbonize without people noticing – 'stealth strategies'. An academic review of eighteen strategies for climate change, by John Wiseman and colleagues, confirms these findings. The study examined nine government-led strategies and nine from non-government sources. The government-led plans, they concluded, were markedly less ambitious and more incremental in tone than the non-government plans. Yet all the strategies had one thing in common: they had not thought through the politics:

> The strategies typically did not go into great detail about how to address social equity or governance aspects of the transition and this is an area for future consideration and development. There is also a lack of detailed game plans within the strategies analysed for mobilising the required level of political leadership and public support for rapid transitions. This remains the most significant gap

in post-carbon economy transition strategies. (Wiseman et al, 2013, p 91)

No country is yet on the right track. Yet times are changing. As I will discuss in the Chapter Six, many countries have committed to radical carbon reductions. They just don't quite know how they will achieve them. In the next two chapters, I will sketch out how it could be done.

SIX

More, and Better, Democracy

The small town of Penrith, on the edge of the Lake District, is in one of the most northerly English parliamentary constituencies. When Rory Stewart, its MP until the 2019 general election, first decided to stand for election in 2010, he spent six weeks walking through its towns, villages, farms and uplands, meeting the people he wanted to represent. So when, as part of a project with the think tank Green Alliance, we invited him to address and hear from a group of fifteen local people making up a citizens' jury on climate change, we weren't surprised that he said yes.

The citizens who met that day were not climate activists. They had been selected to mirror the make-up of the constituency as a whole, in terms of age, social background and political outlook. Before they arrived in the church hall that morning, they didn't even know what subject they had been invited there to discuss. But as conversations about climate got under way, one overriding feeling emerged from participants: confusion.

The people assembled that day had heard about climate change, and were worried about it, but they were confused: if the reports that they were getting, from scientists, from the media and from TV documentaries like David Attenborough's *Climate Change: The Facts*, were right, then why wasn't there more political attention on climate? They couldn't understand why, if it was so serious, government was not taking a lead. They knew that there were things they could do for themselves – like recycling and driving less – but these seemed like insignificant contributions if they were not backed up by a coherent strategy, led by politicians. As one said, "the Government needs to lead by example – everyone from the top down needs to play their part".

These discussions, between citizens and their representative, were a vivid illustration of what has gone wrong in climate politics. As we saw in Chapter Five, politicians have done very little to involve people in climate action, instead hoping that changes could be made through stealth, without impacting on people's lives. This becomes a self-fulfilling prophecy. People are less likely to support climate action if they don't see their politicians leading the way, and offering up a strategy that is as serious as the problem it is designed to address. Climate politics has become a silent standoff, with neither citizens nor representatives willing to make the first move.

But this deeply unhelpful standoff does contain the seeds of a solution. If politicians have the confidence to lead, to see climate action as a social contract between citizens and politicians, then, our Penrith participants told us, they are likely to be supported. The conversation in that church hall gave both sides the chance to explore what a meaningful response to the climate crisis might look like. What lessons might it provide for the democratic process?

As I set out in Chapter One, I think that what is needed is *more* democracy, not *less*. I have shown that politicians do not feel under pressure to act decisively. But I have also argued that this is not because people are actively choosing to worsen the climate crisis. Instead, it is because not enough has been done to explore what people – what voters – really think, when given the opportunity, and responsibility, to consider our climate future. I am, in essence, arguing that democracy can handle this – but not without significant change.

I will suggest three linked ways in which politics could be improved, in order to develop a meaningful political response to climate: first, practical steps to improve the dialogue between citizens and the state – some of which are, encouragingly, already underway; second, a more explicit, purposeful political story of transition, or transformation – to a different economy, and different social arrangements, that will allow us to thrive without carbon emissions; and third, a deceptively simple suggestion: an appeal to the heart.

Before we come to this, though, it is worth taking a step back, and asking a more profound question: what, actually, is a politician's job? What does it mean to represent a constituency of people or a place? The conversations in Penrith illustrate how crucial these questions are. Politicians' understanding of their representative role is fundamental to thinking through what happens next on climate.

Rethinking representation

It sounds obvious: a politician's job is to represent people. But it is surprisingly hard to pin down exactly what representation means. From the Ancient Greeks onwards, philosophers and politicians have argued over how a representative should carry out their job of representing citizens. Should they just act as a mouthpiece, hearing from people about what their needs are and what they want parliament to do, and channelling these needs and views into parliament as best they can? Or can politicians claim to know better than the people they represent, by virtue of their position, experience or education? In which case, they may listen to people's opinions and experiences but will feel free, ultimately, to say what they think is best.

Different politicians, and different political traditions, have different answers to these questions, and in reality, most politicians will do a bit of both. In the UK system that I studied, there is a particularly strong attachment to place. One MP I interviewed described "the incredibly real emotional pressure you feel from your constituency". She used metaphors of family to explain this:

'It is really unbelievably strong. It's almost like being a parent. It's got the sort of joys and terror associated with that type of emotional connection. And so you see everything through the prism of how it will go down locally, and if you're properly connected you will in a sense become your constituency, and you will walk in time to it.'

Yet other political traditions have a very different outlook. Sweden, for example, has a national list system in which national politicians are not linked to any particular local area.

The political theorist Michael Saward cuts through this debate about whether politicians lead or follow (Saward 2010). He says that it is wrong to see representation as a static fact. It is instead, he argued, a negotiated claim, between a representative and the people who elected her or him. Winning an election is necessary but not sufficient. Representation should be seen as a process of 'claims-making', in which the politician makes claims which are then accepted, rejected or ignored by the electorate. In short, representation is a dialogue. When an MP campaigns against a hospital closure, for example, they are, in effect, saying "I am campaigning for local health services and this makes me a worthy representative of this area". Saward calls this a 'representative claim' (Saward, 2010).

Making a representative claim is sometimes straightforward. A politician can simply say that it is in the interests of the people they represent to have a local hospital, and this claim will be widely accepted. Making claims based around local schools, jobs and infrastructure are all obvious ways in which politicians can demonstrate how they represent their local area. On climate, though – a complex, global issue with no clear beginning or end – making a claim takes some work. As one politician told me, "it's nice to have very precise issues that you can get clear wins on". In contrast, with climate:

'Yes, it affects your constituency. Your constituency's a tiny part of the planet; you're not the worst affected part of it. It's just too big, I think, and overwhelming, and I think a lot of politicians like to have campaigns that they can win. Then they can do that press release, I campaigned for x, and I got x, and I've delivered for you. And you can't say "I've campaigned to stop climate change. And now climate change is fixed, and I've delivered for you." That's never going to be a press release that anyone's going to put out.'

To make a representative claim on climate change, then, politicians need to slice and package, to parcel up a global issue into something that is meaningful and manageable to the electorate. Among the politicians I worked with, I saw four different types of claim. First, some made what I called a 'cosmopolitan claim', which emerges directly from climate science. A cosmopolitan claim frames climate change as a global problem to which a global solution is proposed. Politicians argue that it is in the interests of the global community to take action. As one interviewee told me, "a lot of the impacts of climate change are going to hit other places before they hit here. [My constituency] is not likely to be one of the first places to be hit particularly badly. So what? I just happen to be here."

This claim has the advantage of acknowledging the global dimensions of the problem. It is, essentially, the claim used by climate activists. Act now. Yet it has limited appeal, as another explained, given that many people "fundamentally care about themselves, their environment, their friends, their local space ... We have these sort of massive big things about what will happen in other parts of the world ... and they're like, 'yeah, ok, whatever'." In short, as I discussed in Chapter Four, this claim is often ignored, because people are understandably caught up with daily lives and struggles, or feel powerless, cynical or overwhelmed by the scale of the challenge.

Another strategy is to tailor the claim explicitly to a local setting, saying that action is necessary to prevent local impacts like flooding. This could be called a 'local prevention claim'. One politician representing a flood-prone area told me that he used floods as a way of talking about wider climate impacts. This claim has the advantage that it links a global issue directly to the local area, and allows a politician to talk in terms of the interests of local people. As with the cosmopolitan claim, though, it does not link directly to a case for local action to reduce greenhouse gas emissions.

The most common strategy that interviewees reported using was linking climate change to practical, achievable local actions, particularly economic measures, such as encouraging renewable energy generation, or improving transport infrastructure: a 'co-benefits claim'. This has the obvious advantage of relevance to the local area. As one MP told me, "I'm happy to use an economic argument if that means that more people will come on side ... I change the language to be much, much less extreme." The disadvantage of such a claim, though, is that it may reduce the opportunity to discuss the full implications of climate change, focusing instead on small steps at a local level.

Last, a significant minority of MPs in my sample used an intriguing strategy, which I call the 'surrogate claim'. This approach involves promoting local benefits, like public transport or reduced congestion, with no mention of carbon savings or climate change. In this case, although the politician is privately thinking of a particular strategy in terms of its climate benefits, they deliberately do not mention this, because they think it would backfire. One judged that, if he had mentioned carbon emissions in arguing for a sustainable transport scheme, "there would have been a rolling of eyes and saying, 'oh here he goes again'". This approach – trying to act on climate without people noticing – ties strongly with the 'stealth strategies' approach to climate policy that I described in Chapter Five.

In short, this theory of claims-making explains how politicians can, if they choose, bridge the gap between the complex, mediated, global issue and the links with everyday lives of voters. But it also shows that making a case for climate action is much more complicated than simply pointing out the scientific case. As one interviewee memorably said, "You don't say someone came to my surgery with climate change coming out of their ears". They found it harder to make a claim on climate than other issues, such as supporting local services or providing job opportunities. However, it is possible to build a claim for why they should be advocating action.

This analysis of how representation works also helps in thinking through how political systems could change for the better, and how politicians could be supported to make stronger, more radical claims for climate action. I turn to this next.

Opening up to deliberation

Politicians may seem adept at crafting claims which appeal to the people they represent, but this is often based on a less thorough understanding of those people than you might imagine. Research in the US, for example, has consistently shown that voters are less right-wing than politicians think (Broockman and Skovron, 2013). In the UK, my work showed that politicians have underestimated levels of public concern. A consistent refrain from politicians was that they did not feel that they had a mandate to act radically on climate – despite high levels of public concern being reported in polls. Neither did they have a clear sense of whether policies they might put forward will be supported or opposed. As a result, the tendency is toward caution, at a time when scientific evidence on climate screams for rapid and radical action.

A crucial way to break this impasse is to talk more: to encourage direct, structured conversations between politicians and people they represent. The citizens' jury in Penrith, described in Chapter Five, is one such example of a process. Many local areas in the UK have pledged to hold similar local processes, to develop local climate strategies. Plans are also underway for national 'citizens' assemblies' on climate change in both France and the UK. Deliberative processes like these allow citizens, experts and politicians to meet on equal terms, to learn, debate and agree solutions (Dryzek and Niemeyer, 2008). This is not a substitute for electoral politics, but it is a practical means by which elected politicians can explore the public mandate for action – developing 'representative claims' which people can support.

The experience of the Republic of Ireland shows the potential benefits. In 2016, Irish politicians agreed to establish a citizens' assembly as a way of better involving Irish citizens in the political process. The aim was for the assembly to debate a small number of crucial issues, and provide recommendations back to the Irish Parliament. A Supreme Court Judge, Mary Lafoy, was chosen to chair the assembly. The other 99 members were selected from the electoral register, balanced to mirror the population as a whole, in terms of age, gender, social background and geographical location.

The first issue the assembly debated was the Eighth Amendment of the Constitution of Ireland, which set out the equal right to life of a pregnant woman and her unborn child. This is the amendment that was overturned in the historic 2018 referendum vote, which reinstated women's rights to abortion. While the referendum itself was reported around the world, there was much less focus on the vital process that allowed it to happen: the deliberations and recommendations of the citizens' assembly. Over the course of five meetings, assembly members had heard from and debated with legal and medical experts, religious organizations and many other groups. They then debated and voted on their recommendations, with a significant majority voting for change to the constitution, thus paving the way for the referendum. The rest is history.

The citizens' assembly did not, of course, somehow magic away the strong feelings and deep divisions on this issue in Ireland. It didn't lead to consensus. That wasn't the goal. But it did allow an inclusive debate to take place and provided a route to reform which, it turned out, had the backing of a solid majority of Irish citizens.

Thanks to pressure from Ireland's Green Party, another topic that had been put on the list for the citizens' assembly was climate change. The question the assembly was posed was disarmingly simple: 'How can the state make Ireland a leader in tackling climate change?' Originally, a single weekend's debate had been envisaged. But assembly members wanted more time, so an extra weekend was added. It worked in the same way as the Eighth Amendment debate: the Assembly received 1,200 written submissions, and 21 individuals with particular expertise or experience appeared in person to debate with assembly members.

Remember the prevailing view of the politicians that I spoke to and the research showing that people rarely raise climate change as an issue when asked by politicians or opinion pollsters. The contrast with the results of Ireland's assembly is striking. A huge majority, 97 per cent of citizen participants, said that climate change should be 'at the centre of policymaking in Ireland', (The Citizens' Assembly, 2018, p 5) with an independent body, adequately resourced, overseeing progress. Eighty per cent said they 'would be willing to pay higher taxes on carbon intensive activities' (The Citizens' Assembly, 2018, p 40).

This question on taxation deserves a closer look. There were three key conditions set out for the proposed programme. The first was that the revenues from tax should be spent on 'measures that directly aid the transition to a low carbon and climate resilient Ireland' (The Citizens' Assembly, 2018, p 30), like encouraging the uptake of solar panels and energy efficiency measures, as well as flood defences. Second, the assembly specifically stated that they did not want the poorest households to pay higher taxes, acknowledging the equity dimensions of the issue. And last, they recommended that these taxes should be increased year on year.

Remember the two main criticisms of current climate strategies that I outlined in Chapter Five: first, the feelgood fallacy, promoting the positive while not tackling the fundamentals of the fossil-based economy; second, the tendency for stealth strategies, suggested in the hope that experts could impose policies without people noticing.

Their debate and recommendations on taxation show that the assembly didn't fall into either of these traps. Their recommendation for higher taxes on carbon-intensive activities gets straight to the root of the problem. It acknowledges the need to reduce high carbon activities, not just do more of the good stuff. Suggesting higher taxes – with the proceeds spent on the transition to 'a low carbon and resilient Ireland' – is eye-catching to say the least, the very opposite of a stealth strategy. And their acknowledgement of the social impacts of this transition, with their insistence on making sure that poorer households don't lose out, makes it clear that climate policies are about society, not just technology.

Turning to the specific policies, those that the assembly recommended are, perhaps unsurprisingly, ones that resonate and connect with people. They suggested that: individuals should be allowed to sell power from solar panels back to the electricity grid; renewable energy projects should be owned by local communities; and that public transport, walking and cycling should be prioritized over new roads. On many points, assembly members showed more radicalism than most politicians, saying that 'the State should prioritise the expansion of public transport spending over new road infrastructure spending at a ratio of no less than 2-to-1' (The Citizens' Assembly, 2018, p 6), and recommending taxes on greenhouse gas emissions from agriculture. These are both measures that mainstream politicians have generally thought too risky or bold to consider.

Reading the reports from the participants in the assembly, it is clear to me that citizens valued the chance to learn about and debate these issues. As one wrote:

> I feel lucky to have been present during these discussions however it makes me worry about the rest of the population. They are not educated on it. They are not aware of what needs to be done. They are not aware of how to make changes. We need to educate people make them aware of how and why. The government needs to make it possible. Introduce incentives to help people make these changes. I firmly believe that, if educated, the majority of the people would make these changes. (The Citizens' Assembly, 2018, p 115)

The UK Parliament's citizens' assembly on climate change – which began work in January 2020 – is an even more ambitious undertaking. It is a collaboration between six parliamentary committees, involving politicians with responsibility for policy areas including transport, energy, economic policy and local government. Meanwhile, the French *Convention citoyenne pour le climat* will follow a similar pattern, with 150 citizens deliberating over potential climate policies, including the infamous carbon tax proposals that brought the *gilets jaunes* protesters to the streets in 2018.

The French process is perhaps the most ambitious, because it is not just an advisory body. The government has said that it will be bound by the decisions that citizens come to. Participants will learn, deliberate, reach conclusions and then work with lawyers to form them into legislative proposals. Political scientist Loïc Blondiaux, who is part of the team running the process, is adamant that this approach strengthens representative democracy, rather than replacing it. He says that democracies in their current form are not well placed to tackle climate change:

> We need new forms of democracy. Representative democracy has worked well for two centuries. It has succeeded in responding to the challenges our society has faced. It's no longer capable of doing that. Now, we need to think about supplementing it with participative democracy. (Blondiaux, 2019; my translation)

The crunch issue for all these processes is, of course, the extent to which they influence political decision-making. In June 2019, the Irish government published its climate strategy, regarded as much more ambitious than previous iterations, and accepting many recommendations from the citizens' assembly, though still too cautious for many (Darby, 2019). In the UK, it is too early to say whether its new assembly will result in significant new climate commitments from the government. Much depends, not just on the results of the processes, but also on wider political currents.

A more deliberative approach to politics, though, is not just about big events like citizens' assemblies. It is less about specific processes, and more about a different way of doing politics, as we can see from a US example. Back in 1999, eight electricity companies in Texas held a series of deliberative polls; like an assembly or jury, these provided an opportunity for randomly selected citizens to consider evidence and offer recommendations on a specific issue. Citizens were asked for their views on what energy options they preferred, in order to meet future electricity requirements (Lehr et al, 2003).

There were two clear lessons from the Texas experience. First, people changed their views once they had developed knowledge on the issues at stake. At the end of the process, citizens offered stronger support for renewable energy generation and for making energy use more efficient. Second, the deliberation helped both the companies involved and the legislative state of Texas to back wind energy projects. Following the polls, the Texas legislature implemented a 'renewable portfolio standard' to encourage wind energy. Twenty years later, the state of Texas is by far the highest generator of electricity through wind, compared with other US states.

This is why deliberation is so crucial to climate politics. It is a specific acknowledgement that tackling the climate crisis takes us into uncharted territory. People are concerned about climate, but uncertain about what can be done about it. Meanwhile, politicians worry that they won't have support for climate policies. At its best, deliberation allows for structured dialogue between citizens, experts and politicians, to chart a collaborative way forward. To use the language of representative claims, it allows politicians to test and implement claims that will resonate with the electorate.

A more deliberative approach to politics can also unsettle and confront power. In Chapter Three, I described the disproportionate influence wielded by high-carbon interests. Deliberative processes can confront this, by providing a controlled space within which to discuss and debate, with the power held by citizen participants, not paid lobbyists (Curato et al, 2017). This, in turn, gives those politicians who want to challenge current power structures the evidence and confidence to do so. As one participant in Ireland's citizens' assembly noted, 'being bold and ambitious will be severely weakened and outcomes watered down if we allow vested interests to circumvent the efforts of others' (The Citizens' Assembly, 2018, p 115). Loïc Blondiaux, adviser to the French process, echoes this: 'The citizens involved don't have any vested interests. They will be able to do the job they are given… representative democracy is captured by vested interests and lobbies who have an influence' (Blondiaux, 2019; my translation).

Deliberation, then, is a key ingredient in the better democracy that is needed to respond to climate change. It goes beyond the commissioning of one-off processes, like citizens' assemblies. It is instead an outlook: more painstaking to-and-fro between politicians and the electorate, a way of doing politics that acknowledges that neither side has all the answers, and a deliberate up-ending of existing power relations. It is a social contract between citizen and state, in a constant, and constructive, state of negotiation, in response to the uncertainty and enormity of the climate challenge.

It is tempting to conclude that political action on climate will only be possible with radical reforms to democratic systems, including stricter controls on political funding. There is much evidence, too, to suggest that democracies perform better in more equal societies (Milanović, 2016). For the reasons I have discussed, I agree that efforts to improve the democratic process would lead to better climate politics. But waiting for perfect democratic conditions is not really an option – especially when, as I discussed in Chapter Two, climate change is itself likely to lead to greater political uncertainty. In this book, I have tried to start from where we are, rather than prescribing ideal conditions.

A story of transformation

Another crucial ingredient in an improved democracy is a story of transformation. We saw in Chapter Five how 'stealth strategies' – trying to substitute low-carbon approaches for high-carbon ones, without people noticing – are inherently self-limiting. By definition, they do not make the case for change; they do not tell a story about the transformation that will be needed, if our societies and economies are to shift. Neither do they talk about the other benefits of climate action.

When Democrat Congresswoman Alexandria Ocasio-Cortez burst onto the US political scene in 2018, battling her own party's establishment to do so, she brought with her a startling new approach to climate politics: the Green New Deal, an unashamedly idealistic plan to transition the US to a socially just society with a zero-carbon economy. Alongside an aim for net-zero greenhouse gas emissions and 100 per cent renewable energy, the Green New Deal demands job creation in manufacturing, economic justice for those living in poverty and for minorities, and even universal healthcare, through a ten-year 'national mobilization', a phrase which echoes President Franklin Roosevelt's New Deal in the 1930s.

The Green New Deal is an audacious representative claim. It is, in some respects, a cosmopolitan claim – it stresses the global nature of climate change, and the need to act. But it also claims strong social co-benefits, like jobs and healthcare.

The plan has many critics, not just in the Republican Party but within Ocasio-Cortez's own ranks. House Speaker Nancy Pelosi famously dismissed it as a 'green dream'. But what it offers is a story of transformation: a clear, engaging account of the transition to a zero-carbon future. Compare that to the low-key approach favoured by the UK politicians that were the subject of my research: dressing up climate action in the language of economic policy and market mechanisms to avoid confrontation with colleagues, the electorate or the industries that risk losing out in the shift to a low-carbon economy; or using stealth strategies and deliberately avoiding any mention of climate change.

Sure, the Green New Deal is controversial, and idealistic. That's the point. Although it will fail, it has already succeeded in one important respect: it has catapulted climate change to the top of the US political agenda. Candidates in the race for the Democratic presidential nomination have been forced to articulate their position on climate, instead of hiding behind a wall of silence. The Green New Deal has also put the issue of corporate influence, and fossil lobbying, centre-stage. It is upfront about the contradictions in high-carbon societies.

The Green New Deal is not the only possible story of trans-formation. But its opponents can't simply criticize it. They will need to find their own answer to the climate question. The US debate has drawn attention to a gaping hole in right-wing politics: the confident articulation of a climate strategy. Elements of the Republican Party will continue to deny or dismiss the science. But in the US, the UK and many other countries, the race is on for political conservatives to respond. If you agree with the scientific consensus that rapid action is necessary, but you don't like the strongly social flavour of the Green New Deal, what do you propose in its place?

Even as the UK was mired in Brexit stalemate, a group of 41 Conservative MPs sought to answer this question. In July 2019 they launched a manifesto acknowledging the scale of the climate crisis, and proposing a raft of policies, based around the idea that markets, steered and influenced by government, could deliver climate solutions. Most importantly, they told a story. In their words:

> This is an optimistic document. It argues that tackling the existential threat of environmental breakdown offers our divided country a new national project. It argues that this unifying mission can bring economic regeneration and natural restoration to all parts of the country. (Conservative Environment Network, 2019)

As I have already argued, climate is a more problematic issue for parties of the right than the left, because it is, by definition, a market failure; solutions almost always involve intervention and collective action. This is acknowledged by many on the right, including the authors of the manifesto, who argue that the task is to shape markets to deliver the right outcomes. As the UK takes stock after its third election in five years, it remains to be seen whether this vision will gain traction.

An appeal to the heart

Last, improving democratic conversations on climate change almost certainly means making more of an appeal to the heart as well as the head. The desiccated language of cost-benefit analysis and gross domestic product has its uses, but its appeal is limited and it accentuates the divide between experts and public, rather than breaking it down. Yet this is how politicians have described climate change.

When I analyzed the language used by politicians to talk about climate, I uncovered a marked tendency to use scientific and technical language, and a startling lack of discussion of people, families and society. Words within these groups were statistically underrepresented in debates about climate change, compared with everyday speech. At first, I couldn't quite believe these results. I thought perhaps there was a natural difference between politicians' speech and the database of everyday speech that I was comparing it with. So I ran the same analysis for politicians' speeches about economic policy. To my surprise, politicians were six times more likely to use words associated with people and family when they were talking about the Budget, than when they were talking about climate change. They presented finance and budgets in a much more people-centred way. Climate language was technical, not social or emotional (Willis, 2017).

In recent years it has been the populist right, reliably opposed to climate action, who have best articulated emotion in politics, positioning themselves as supporting 'the people' and opposing 'experts' and 'elites'. As I discussed in Chapter Two, this tendency is deeply problematic for climate politics. The more that climate is seen as an elite, expert, technocratic agenda, the more likely it is to be rejected by voters who see it as a distraction from their interests and needs.

Like me, Alex Evans spent many years advising government on climate change. He worked as a special adviser to two British cabinet ministers and as an adviser in the UN Secretary-General's office. In recent years, he has written compellingly about the role of myth, emotion and psychology in climate action. As he writes:

I used to be sure that with science on our side, policy change would naturally follow. If only. Instead, we haven't even begun to reduce global emissions. Why? In a nutshell, because opponents of climate action have too often had the better stories, and stories always beat data. (Evans, 2019)

Yet there is no reason why climate should be framed just as an expert-led technical agenda, or why right-wing populists should own all appeals to emotion. When I read Claire Ainsley's book on the views of Britain's working class, I was surprised how closely her conclusions about political strategy chimed with my own. She advocates an agenda based on public values of family, fairness, hard work and decency, with an active role for the state, and a view to the long term (Ainsley, 2018). This is exactly how climate action should be framed, too. As Alex Evans writes, our task is to tell 'stories that make us think in terms of a larger us, a longer now, and a better good life. Stories that don't flinch from telling us how close to the edge we are on issues, like climate change, but also that give us hope' (Evans, 2019).

No story is more powerful than the story of love. In a startling and moving recent essay, climate justice activist Mary Heglar wrote that love, not guilt, fear, panic or anger, was her overriding response to climate change. Love for 'this beautiful, mysterious, complicated planet' and for family and friends. Heglar ends her essay with these words:

> It's a love that can also – even in the teeth of these most insurmountable odds – give me hope. If I'm brave enough to accept it. I've seen her looking back at me in the eyes of some of the bravest climate justice warriors I have ever met, and I can feel that tickling tingle of 'maybe, just maybe, we'll be okay.'
>
> A love like that doesn't seek peace, or even vengeance. She seeks justice. And she's strong enough, ferocious enough, brave enough to burn this bitch to the ground. (Heglar, 2019)

The political principles I have outlined in this chapter – deliberation, debate, transformation and, yes, love – point to a very different model of climate politics than that which has been dominant for thirty years. And yet we are still in the realm of the abstract. In Chapter Seven, I will outline what this might mean in terms of political practice and policies.

SEVEN

A Strategy for the Climate Emergency

Twenty years of climate policy have seen some successes, but global emissions are still rising. No country yet has a comprehensive strategy for carbon reduction that meets what the scientific community say is required. And climate impacts themselves could result in worsening conditions for democratic governments.

We saw in Chapter Five that climate strategies initiated by countries and local areas had been limited, for two reasons: first, they focused on encouraging positive action, like renewable energy, or greener products, but did not discourage high-carbon activities, like fossil fuel extraction, or car use. I called this the 'feelgood fallacy'. There are plenty of schemes to encourage electric vehicles, and some countries charge less tax for smaller cars, but the incentives are not large. No country has yet put significant measures in place to discourage people from buying cars with large petrol or diesel engines (Wappelhorst, 2018). There are no restrictions on advertising cars, either. Similarly, while there are incentives for renewable energy in most European countries, there are still few restrictions on fossil energy. The UK has generous tax breaks for oil and gas exploration in the North Sea, and these have increased in recent years, despite the country's ambitious carbon targets.

The second difficulty is that much climate action has been carried out without the involvement or meaningful consent of people. The shift from a high-carbon society to a zero-carbon society is not just technological. It requires social and cultural change too. Yet as we saw in Chapter Six, politicians and policymakers are often tempted to adopt 'stealth strategies', rather than making an upfront case for change. Like the politician who argued for a local transport scheme, without mentioning carbon reduction, they feel it's best not to open debate about climate change.

Given this track record, many critics say that our politics is fundamentally broken. Spurred on by Naomi Klein's assertion that 'this changes everything' (Klein, 2015), many argue that nothing less than an end to capitalist economic systems will solve climate change: 'system change not climate change', as the protest placards put it. In a similar vein, economist Tim Jackson (2017) argues that economic growth, the goal of nearly every government, is incompatible with climate stability.

These are powerful arguments – not least because the current political and economic system has demonstrably failed on climate. Yet so far, these debates have largely remained outside the sphere of formal politics. None of the MPs in my research were advocating an end to capitalist economic systems. This could be because they are part of a politics that sustains, and is sustained by, the current system – maybe they are too embedded to see the bigger picture, or they feel that they stand to lose from a more wholesale system critique (Machin, 2013; Mouffe, 2000). Whatever the explanation, 'system change' is too big a leap for many politicians.

This is illustrated well by one pithy exchange I had, with a veteran ex-minister, during my research. We were talking about measures he had supported to increase cycling and the take-up of renewable energy. He told me it was important to start with simple, achievable steps. I had been murmuring my agreement up until this point. But then I started to challenge him a little. I pointed out the urgency of the climate crisis and the need to reduce emissions. I said that this was a situation that hadn't been faced before – that there had been a relatively stable climate for all of human history. His response was emphatic. He said, "the argument you've just made, that we're in a qualitatively different situation than we've ever been in history, in my opinion doesn't help at all". He didn't disagree with me – but he wanted me to see it from his point of view, and understand what he could do, in the here and now.

Radicalism or pragmatism?

The conversation I described above will be painfully familiar to anyone who has worked in politics or campaigning. This is the fundamental dilemma of influencing: it often seems to be a choice between pragmatism and radicalism. Pragmatic influencers start from the current situation, understand who has the power and what motivates them, and work out how to make progress within these constraints. They are more likely to take the inside track: small meetings, consultations, advisory groups. Radical influencers, instead, start from their ideal end point, stating a clear and unequivocal goal, working through public campaigns, activism and media, as well as speaking directly to politicians. There's a risk to either strategy. Too pragmatic, and you risk being coopted; too radical and you risk being marginalized.

I've always leaned toward pragmatism. I started my career working for politicians in the European Parliament; I saw firsthand the negotiations and compromises that are part of political life. Green Alliance, who I worked for over many years, were so proud of their close relationship with government and politicians that they named their magazine *Inside Track*. Everyone who has worked at Green Alliance could give you a few good stories about how a particular tax measure, environmental standard or policy proposal emerged from a carefully designed process of research and influencing. I remember a tense wait, back in the late 1990s, when Gordon Brown stood up to give one of his first Budget speeches as Chancellor. We were hoping he would announce the business tax on carbon that we had spent many months working on. He did.

Yet even these small steps forward would probably not have been possible without campaigns and activism from more radical groups like Greenpeace and Friends of the Earth. They purposefully make life difficult for government, using their supporters and media tactics to increase the pressure, allowing organizations like Green Alliance to appear at just the right moment, and suggest ways through.

When I look back on my experience in the UK over the past decade, though, I think we got the balance wrong. We were too patient, too understanding, too willing to give successive governments the benefit of the doubt. Maybe we were lulled into a false sense of security by the Climate Change Act, which set ambitious targets, if not the means to achieve them. When I ran a focus group discussion with people working for environmental groups – a mix of the radical and the incremental – participants explained how their strategy had been to present climate action as an economically beneficial strategy. One likened their work to corporate consultancies, saying that they would "pretend that we're McKinsey and EY … it's been really helpful in winning the overall argument, shifting how climate change is perceived".

Political scientists refer to this process of crafting messages as 'framing', when messages are shaped to fit a sense of what is achievable, without confronting dominant norms or assumptions. It relates to the human urge to fit in, that I discussed in Chapter Four – not wanting to be a 'climate change zealot', as one politician put it. I don't deny this can be a useful tactic. But we relied on it too much. One politician I interviewed, a veteran climate campaigner, criticized experts, including people from environmental groups and the research community. She said that visitors to Parliament, keen to appear reasonable and rational, were telling MPs what they wanted to hear. "It made me so angry," she said. "It's something about this place, I think, once you get inside one of those Committee rooms."

When Greta Thunberg burst onto the scene in 2018, she made many environmentalists, including me, sit up and think.

Thunberg does not equivocate. She knows that what she is saying is deeply uncomfortable to her audience, and she says it anyway.

What we need, in response to the climate emergency, is a steadfast radicalism in terms of aims coupled with an unfailing focus on the process of transition or transformation. We need to link the wholesale critique of our economic and political system to the equally crucial question of what should happen, today. How can we build out from our current base? What should we be asking of our politicians? What does a meaningful response looks like – one that acknowledges the constraints of our current system, but does not compromise on the changes that are needed?

That meaningful response is not a list of technological or policy responses to climate change. There is plenty of evidence about what works from a technical perspective. But much less thought has gone into how to win democratic support, acknowledging people's values and engaging them, not attempting to bypass them. I will now begin to sketch out a climate agenda, based on the assumption that democracy matters. I do not go into detail on particular policies for particular sectors: as I have discussed throughout this book, there is no shortage of proposals. The gap, instead, is the question of how such proposals could be brought about, through the democratic process.

I suggest that such an agenda involves, first, an upfront acknowledgement of the climate crisis, and a recognition that responding to climate change has huge significance for the way in which each country, and local area, is governed. Second, it requires a clear strategy, with democratic oversight through deliberative processes such as citizens' assemblies. Such a plan needs to consider how to divest and move away from high-carbon production and consumption, not just promote low-carbon alternatives. Next, strategies should be considered specifically in the context of public engagement: do the proposals engage, motivate and build the mandate for more ambitious action? The local level is almost certainly the right location for many such strategies. Finally, national and local strategies must acknowledge the global context.

I end with a climate strategy checklist – a deliberately simple, yet I hope comprehensive, list of ten areas that any country's climate strategy should address. I can't think of a single country, yet, that scores ten out of ten.

An upfront acknowledgement of the climate crisis

In November 2018, the City of Bristol started something huge. Led by a Green Party councillor, Carla Denyer, the city council declared a 'climate emergency'. Acknowledging the impacts of climate change, councillors pledged to make the city of Bristol carbon neutral by 2030. Councils around the country started to do the same: as Denyer said, the Bristol declaration 'is triggering copycat motions all over the UK, and honestly I've never been happier to have my homework copied!' (Denyer, 2018).

By mid-2019, spurred on by striking schoolchildren and climate protests, over 220 UK local authorities, as well as the devolved nations of Scotland and Wales, had copied Denyer's homework and issued similar declarations (Climate Emergency, 2019). Then, in May 2019, the UK Parliament itself declared a climate emergency, the first of four countries – the UK, France, Canada and Ireland – to do so, as of mid-2019.

This development is hugely encouraging. The first, and most fundamental, task of an honest climate politics is for politicians to speak openly about the significance of climate change. Unflinching honesty is needed, if politicians are to face up to the fragility of human society, and its dependence on increasingly volatile earth systems. This sounds simple, but as my research has shown, it is a very difficult thing to do. As we have seen, until recently, politicians had tended to present climate change as a less significant, more manageable issue, playing down the scientific evidence and presenting a straight-forward account of climate change as something that can be 'solved' (Willis, 2017). These declarations might signal the end of the stultifying climate silence of recent years.

And yet climate emergency declarations have not, so far at least, resulted in commensurate changes to policy and strategy. Politicians may have acknowledged the climate crisis, but they have also compartmentalized it. There has been little discussion of the far-reaching implications of the declarations that have been made. In Canada, the day after the declaration of a climate emergency, Justin Trudeau's government approved plans for the Trans Mountain pipeline expansion, shipping oil across the country. Even in trailblazing Bristol, the expansion of the local airport appears to have been deemed compatible with a climate emergency: the mayor supports both.

What seems to be happening is an official version of the 'socially organized denial' of climate change that we saw in Chapter Three: a generalized acceptance of the issue, evidenced by expressions of concern, but accompanied by a reluctance to open up to a full discussion of the implications of this for social or political life. That is not to say that the emergency declarations have been worthless, or cynical – it's just that they are only a start.

A clear strategy with democratic oversight

The UK's Climate Change Act is an impressively clear strategy for tackling climate change, up to a point. It sets a long-term target, and interim five-yearly 'carbon budgets'. It was amended in 2015 to increase ambition, to set a goal of net-zero emissions by 2050. There is independent oversight by the Committee on Climate Change, who report progress to Parliament each year, and a requirement for the government to respond. So far, so transparent. There is much that other countries could learn from.

And yet, as the Committee itself recently made clear, the UK is not on track to meet its targets. There are some gaping holes in the Climate Change Act. It is not clear how government departments with responsibility for transport, housing, local government and other issues are required to contribute to decarbonization. They have no formal targets or responsibilities under the Act. Neither do local areas have targets, duties or resources to act on climate. There was a plan to strengthen the Act with further measures to divide up responsibilities in this way, giving explicit duties to all government ministers and all local areas to meet carbon targets (HM Government, 2009). But the 2010 general election got in the way, a new government arrived, with different ideas, and the final parts of the plan were forgotten.

It sounds almost too obvious to state, but a strategy for the climate emergency should assign clear responsibility to each government department and agency, and to governments at local and regional level. The Department for Transport, for example, would be responsible for reducing emissions from transport, in line with the overall budget. It could not be assumed – as it has been in discussions over aviation, for example – that increases in transport emissions would be offset by reductions elsewhere in the economy.

At local level, local politicians should be required to set local plans, which are responsive to the needs and conditions of their area. Andrew Cooper, a local politician in Yorkshire, has come up with an intriguing plan for how this could work. In Chapter Two, I outlined how, under the Paris Agreement, each country was responsible for developing a national plan, called a Nationally Determined Contribution (NDC). Cooper has suggested that local areas could develop a Locally Determined Contribution (LDC), which sets carbon reduction targets and timescales, following an agreed methodology that allows comparability across local areas. Local areas would negotiate with central government what contribution they would make, and what resources and powers they would have to achieve their strategy (Cooper, 2018).

Transparent data and measurement are central to any climate strategy. The cliché that you can't manage what you can't measure is a cliché because it is true. A comprehensive response to climate change involves being very clear about a baseline (the amount of carbon being emitted at the start), targets, and responsibility for those targets.

This all sounds very technical and procedural. Yet there is a deeply political point buried here. At the moment, it is just too easy to hide behind the numbers.

That's exactly what the Scottish Government is doing. Their 'programme for government' proudly declares that Scotland will lead the way on climate. In the words of First Minister Nicola Sturgeon, 'In April this year, I acknowledged that we faced a climate emergency. … We are leading the world in setting challenging targets, but we must also redouble our efforts to meet them' (Scottish Government, 2019, p 3). There are, it is true, some admirable commitments to improve public transport, invest in environmental technologies and decarbonize home heating. But there is also a truly remarkable pledge. The plan is to 'establish a new Net Zero Solution Centre, enabling the North Sea to become the first net zero hydrocarbon basin in the world' (Scottish Government, 2019, p 50). The Scottish Government is claiming that extracting oil and gas from the North Sea will not result in additional carbon emissions.

How can they claim this? I am not sure. They offer no calculations, or definitions. It appears to be a mixture of three things: first, reduced emissions from the extraction process itself; second, capturing the carbon and storing it underground; and third, not counting the emissions from the oil and gas itself, if it is burned elsewhere. A net zero hydrocarbon basin, it seems, is a product of creative accounting.

The Scottish Government may be committed to tackling climate change, but it is also still committed to getting fossil fuels out the ground. If there were better figures behind their climate strategy, this contradiction would be obvious.

The reason that few areas have a carefully worked through budget of this sort is that it would expose the feelgood fallacy for what it is. It would no longer be enough to install some solar panels and fund some greentech entrepreneurs. A good carbon budget process will tell you if your efforts to support renewable energy have been cancelled out, in carbon terms, by people buying bigger cars or by generating more energy from fossil fuels. If the numbers are worked out, it is harder to hide.

As I discussed in Chapter Six, climate strategy at both national and local level would also benefit greatly from ongoing deliberation with citizens. For the UK, for example, this could link to the carbon budgeting process, with input from citizen deliberation informing each budget.

A transition away from fossil fuels and high-carbon systems

A strategy for the climate emergency also has to transition away from fossil fuels. If there is any chance of stabilizing global temperatures at 1.5°C or 2°C, which the world signed up to through the Paris Agreement, most known fossil fuel reserves will have to be left untapped (McGlade and Ekins, 2015). Neither can any new fossil-fuel power plants be built. In fact, some existing plants will have to be retired early (Tong et al, 2019). The 'keep it in the ground' slogan seen on activists' banners is an accurate representation of the best scientific evidence.

This means that any climate strategy, for any country or local area, must be explicit about the need to transition away from fossil power. Until now, most climate strategies have been silent on this point. They have focused on increasing renewable energy, rather than decreasing fossil fuels. The UK's carbon plans say nothing about continued extraction of oil and gas from the North Sea. Canada is developing new gas pipelines. India and China are both still building coal-fired power plants. Yet any meaningful strategy must outline a plan for transitioning away from fossil fuel extraction and use.

The only leeway on this point, if the Paris commitments are to be met, is to use so-called 'negative emissions technologies' (NETs), sometimes called geoengineering, which I discussed briefly in Chapter Two. These are a range of different technologies which remove carbon from the atmosphere, through biological means like tree planting or land management, or technical methods like direct air capture (machines that suck carbon out of the air). These technologies are not yet proven at scale, but will be needed to reach net-zero carbon emissions.

These technologies will certainly have a role to play. But there is also a real danger that relying on them or on the promises that they offer even though they are unproven at scale, as we are currently doing, will distract time, attention and resources away from carbon reduction strategies. The best way of avoiding this, again, is through an unflinching honesty and transparency, treating NETs as separate to emissions reduction, and separating them out both in accounting and in policy (McLaren et al, 2019).

Doing all this will almost certainly require changes in the way that government listens to interest groups. As I outlined in Chapter Three, those with vested interests in fossil fuels – not only companies but also trades unions – have a huge amount of influence in politics. When it comes to influencing policy, there is a particular issue with trade associations, which represent particular company sectors. Studies have shown that they lobby against legislation which would affect their members, even if it might be in their long-term interest. The system of representation through trade associations also allows individual companies to present a relatively progressive public face, while supporting more regressive lobbying behind the scenes through their association (Caulkin and Collins, 2003).

A number of changes would help to reduce the influence of the fossil fuel lobby. The first would be an acknowledgement of the issue, with greater transparency about the role played by different companies in meeting ministers, seconding staff into government, and so on. A ban on political funding from fossil fuel companies would be another possibility.

Another change would be for government to seek out emerging low-carbon companies, and support them to advise on legislation. For example, in my work on energy governance, we have developed proposals an Energy Transformation Commission, which would oversee the transformation of the energy system (Willis et al, 2019a). In doing so, it would actively engage a wider constituency, including new entrants to the sector and consumer, environment and other interest groups. Another change would be to include greater citizen deliberation in government policy and strategy, as I discussed in Chapter Six.

Policies that matter to people

Recently, I presented my research on politicians at a Royal Society meeting. It was a gathering of climate experts from business, civil society and academia. I was struck that, among the sixty or more people attending, there were only two of us whose talks included discussion of whether and how the climate policies they were proposing would garner democratic support. The assumption that many seemed to be working to was that it was the role of experts to design technically optimal policies, which would then be handed to the politicians to 'sell in' to voters.

This seemed to me to be the wrong way round. I argued that it was both legitimate and necessary to ask of every climate policy, 'does this build public support?' It was generally accepted, at that meeting, that technical feasibility and efficiency were sensible criteria by which to judge policy proposals such as carbon taxes, investment support mechanisms and legislated standards for vehicles or other products. My suggested addition to this list certainly raised a few eyebrows, yet it is a conclusion that emerges very clearly from my research. Policies should be meaningful, in two senses: they should provide a meaningful, material contribution to carbon reduction; and they should be meaningful to people and communities.

The public opinion expert Deborah Mattinson talks about 'symbolic policies', practical measures which encapsulate a vision, or story, about a wider change. Such policies would have tangible impacts, and would also raise the political profile of climate action. They would galvanize citizens and send businesses a clear message that they could base investment decisions on.

The archetypal example of a symbolic policy in the UK is the 'right to buy': a policy introduced by a Conservative government in the 1980s, offering people who were renting a house owned by the local government, the chance to buy it at a reduced rate. The symbolism of this policy was huge. It told a wider story about aspiration, property ownership, self-sufficiency and a smaller state. One specific policy encapsulated the Conservative vision.

A more recent symbolic policy in the UK – and one that is not helpful from a climate point of view – was the price cap on energy, proposed by the Labour Party in 2013, and eventually adopted by the Conservative government (Vaughan, 2017). It worked because it was a simple pledge, which told a story about big companies reaping excess profits, and about standing up for 'hard-working families'. It doesn't actually solve the problems of an outdated system of energy regulation, but that's another story (Willis et al, 2019b).

What would symbolic policies for climate look like? The answer would depend, of course, on the ideological orientation of the government introducing them. Possibilities could include a right to generate and sell renewable energy, at home or in the community; a ban on advertising petrol and diesel cars; free bus travel funded by a levy on aviation; a green jobs guarantee for communities transitioning away from fossil fuel-based jobs, as US Democrats have proposed as part of the Green New Deal package; or even personal carbon allowances, which could be traded between people.

Symbolic policies don't work by themselves. The point is to signal and engage citizens in a wider vision and project. For transport, two symbolic policies – a ban on advertising petrol and diesel cars, and free bus travel, funded by an aviation levy – would be part of a wider transport strategy that paved the way for a transition to electric vehicles, and more investment in public transport, walking and cycling. The symbolic policies are a way into telling a wider story, but the strategy as a whole has to be consistent.

A clear message from the citizens' juries I described in Chapter Six was that people are willing to support far more ambitious proposals than you might expect – but that they need to know that there is a clear strategy in place. In the words of public opinion expert Lucy Bush, who helped to run the juries, "Without any overarching strategy or cross-sector commitment to climate action in place, people are demotivated to do anything personally – they feel their contribution would be pointless" (Buck, 2019).

Take personal carbon allowances. This idea has been around for quite a while. Under this system, people would be each be given an equal allowance of carbon emissions, decreasing year on year, to 'spend' on transport, flights, home energy, and so on. If they wanted to exceed their allowance, they would have to buy extra allowances. This would certainly be an eye-catching symbolic policy, and previous research has indicated that people understand and support the approach (Wallace et al, 2010). But it would only work if people saw why the policy was being introduced. They would want to know how it fitted into a government-led climate strategy; and they would want to know that businesses were being asked to do their bit as well. Above all, they would need to have low-carbon alternatives offered to them. Penalizing someone for driving a car if there is no public transport alternative to get to work is never going to be popular.

Linking to the international effort

Finally, any national climate strategy must be put in an international context. An individual country's strategy matters for two reasons: first, because of the direct reductions in emissions; second, because of the example it sets, or doesn't set, internationally. I am staggered by how often I hear the 'but what about China?' question posed. Boris Johnson recently used his national newspaper column to declare that the UK was doing just fine on climate, and advised Extinction Rebellion to 'take your pink boat to China instead' (Johnson, 2019). Yet climate action is a virtuous circle. It's the same for whole countries as it is for individuals: the more you see others making significant strides, the more you are likely to do so yourself.

The strategy I have outlined is based on my experience in the UK. Many of the points are generalizable, but some are not; every country's strategy depends on circumstances. For countries in the global south, with low levels of emissions and high poverty levels, the emphasis will obviously be different. Under the Paris Agreement, there are measures in place to support these countries with climate finance through the Green Climate Fund – though the realities don't tend to match the promises (Schatelek et al, 2016).

Countries whose economies depend overwhelmingly on fossil fuel extraction, like Saudi Arabia, Venezuela and many others, are a different case again. So far, their strategy in international climate negotiations has been to delay and oppose. Private companies like Exxon, Shell and BP are rightly coming under pressure from activists, but it is worth remembering that three out of the five biggest carbon-emitting companies are actually state-owned: Saudi Aramco, Gazprom and the National Iranian Oil Company (Taylor and Watts, 2019). Success at a global level will mean these countries shifting away from their main source of income – a daunting prospect for poorer countries with heavy reliance on extractive industries. As researcher Chris Armstrong (Armstrong, 2019, p 15) argues, 'assistance for poor exporting countries is not only politically urgent if extraction is to be limited. It is also required by global justice'.

A striking proposal put forward by Peter Newell and Andrew Simms (2019) is a 'fossil fuel non-proliferation treaty'.

Modelled on the successful nuclear non-proliferation treaty of the 1960s, such a treaty would have three elements: first, a pledge to halt new extraction projects; second, to coordinate and manage the decline of fossil fuel infrastructures, covering both supply and demand; and third, to resource a transition to non-fossil energy, with specific support for those countries which had left fossil fuels in the ground.

A climate strategy checklist

As this discussion shows, climate strategy is either staggeringly simple or incredibly complex, depending on how you look at it. Staggeringly simple, because there are really only four ingredients: stop digging up fossil fuels, and stop using fossil fuels, as fast as possible; manage land to maximize the storage of greenhouse gases; and prepare for the climate impacts that are already baked in. Incredibly complex, because, as I outlined in Chapter Three, fossil fuels and emissions of greenhouse gases are woven in to our daily lives, and into the structure of our societies, economies and political systems.

Despite this complexity, there are some relatively straight-forward questions that any national climate strategy needs to address head on, irrespective of the social, economic or political circumstances of that country, and irrespective of political inclinations. I think there are ten questions that need to be answered:

1. Is there a clearly stated, long-term target, compatible with climate science and international responsibilities, written into law?
2. Is there vocal political leadership on climate change, confident narratives, and a healthy debate about strategy?
3. Are citizens engaged, both through democratic means (voting, deliberative processes like citizens' assemblies, wider engagement and consultation) and through the policies themselves?

4. Is there a plan to achieve this target over a clear timeframe (including the near term), distributed across different parts of government, including all ministries and government organizations, including local or state-level government?
5. Is there independent measurement, verification and scrutiny?
6. Does the strategy cover all the crucial sectors, i.e. transport (including aviation and shipping); power generation; housing and buildings; consumption; industry; finance; land use and agriculture; climate impacts and resilience?
7. Is there a transparent and measurable process for the phasing out of fossil fuel exploration, extraction and use?
8. Are the distributional consequences of the plan being addressed – in terms of protection for individuals; different social groups; job opportunities; and local areas?
9. Do financial flows – government funds, the regulations governing private capital and trade, and development aid – support climate strategy and provide the required investment?
10. Is there a clear strategy for negative emissions, with separate targets and policy for greenhouse gas removal technologies?

These ten areas form the bedrock of any comprehensive climate strategy. The UK, as a self-proclaimed world leader, can say a categorical 'yes' to points 1 and 5: it has a clearly stated target and its plan has independent scrutiny by the Committee on Climate Change. All the rest are work in progress.

If climate strategies do not address these points, it is not because of a lack of understanding, or a lack of evidence. As I have discussed throughout this book, it is because of power: political power, cultural power and economic power. Look back through the list, and think about the powerful people and organizations who feel that they would lose out, in terms of power, money and status, from each of the measures. Think about how they use their influence to delay, obfuscate and oppose. And yet there are many more people and organizations who would support action – and who are staying quiet.

This puts a heavy responsibility on those of us who understand this, and want action on the climate crisis, to hold our government to account. In the final chapter, I take a look at what each of us can do, as individuals.

EIGHT

The Personal Is Political: How To Be a Good Climate Citizen

In the previous two chapters, I sketched out a political agenda for climate: one which acknowledges the severity of the state we are in, while also understanding that the only starting point we have is our current reality. Immense political shifts are needed. The end point has to be a society, and a politics, that has changed beyond recognition – but we are starting with what we have. I have put the case for a different way of doing politics – with more democracy, not less.

In this final chapter, I want to offer some thoughts on what each of us can do, as citizens. We need politicians to lead – but they can't govern in a vacuum. How can people concerned about climate change prod, encourage and challenge politicians and others? In other words, how can you and I be good climate citizens?

Mike Berners-Lee, a friend and colleague and author of some brilliant books on climate, described to me a one-way gate. He says that once you know about the significance of climate change, you have a responsibility to respond. Once you know, you can't forget. There's no way back through that gate. But climate is a big issue, and you are one person. It is hard to know what impact you can have.

When climate hits the headlines, there are invariably a deluge of articles listing all the things people should do differently. They normally involve telling people to switch lights off or use their car less. Recently, there's been more about travel by plane, and diets – and it's certainly true that eating less meat and dairy is good for climate as well as health. But these lists often miss the point. As I discussed in Chapter Three, we are part of a high-carbon system: consumption should not be seen as an individualized pursuit. It's not just what you can do as a consumer.

My research, and my experience of working with politicians, provides a different list. It's not that personal action is not important. But what we consume (or avoid) is only one aspect of who we are. So here are some imperfect, practical, political suggestions.

Speak out

If you're thinking about climate change, talk about it too. Remember the concept of 'socially organized denial', the idea that people accept that climate change is happening but just don't talk about it? There's a way to break through that – speaking out. Whenever you're talking to someone and you think about climate, whether at work or elsewhere, vocalize that thought. It's not an easy thing to do, because, as we explored in Chapter Four, by naming climate change you are saying a lot of difficult things about how we live our lives. In the fable of the emperor's new clothes, it is not a coincidence that it was a child who pointed out that the emperor was, in fact, naked. As Greta Thunberg has demonstrated admirably, children have that enviable ability to ignore social convention and say what they see. We could all learn from that. You won't always be thanked for it, but your courage will be noticed.

You can talk to politicians. National leaders, local councillors – any and all elected representatives. Tell them you are worried about climate change, and ask them what they think. The suggestions in the following section will help you to think about how that conversation might go. The charity Hope For The Future has a brilliant set of resources on its website to help you through the process of asking for, and planning, a meeting.

But don't stop with politicians. You could raise it at your workplace, talk to your friends, make it clear on social media. In short: fight socially organized denial. Last year, I took a deep breath and chatted to the parents on the touchline when my son was playing football. It was a really positive conversation. Since then, I've tried to include climate in a lot of general chat. Or rather, I have stopped censoring myself from speaking about it.

The UK charity Climate Outreach has been encouraging people to have these conversations, and researching the impacts. They worked with volunteers who offered to start up conversations, with strangers, family members, acquaintances and work colleagues, and to report back on their experiences. Though it was sometimes hard to start with, participants were glad they had done it. As one said, 'Talking about it breaks down the isolated feeling, and makes me feel more supported to take action' (Webster, 2019a, p5). This confirms research which suggests that taking action on climate is good for you: it helps overcome feelings of helplessness or grief that may emerge from contemplating something so all-consuming. As Robin Webster, who works for Climate Outreach, told me,

'We've found that people can cope with knowledge of climate impacts, and the strong emotion it prompts, if they are coupled with concrete actions they can take. I've seen how Extinction Rebellion, for example, gives people stuck in despair somewhere to go and something to do, quite fundamentally changing their mindset and attitude, by creating a community to be a part of.' (Webster, 2019b)

Once you've decided to talk about climate, what do you say? It feels difficult, and it is important to acknowledge that. The distant language of scientific reports can mask the reality, but once you think about it in terms of the daily lives of Earth's inhabitants (both human and non-human), it is a shock.

The impacts that are already upon us – wildfires, floods, hurricanes, droughts – are set to increase in number and severity. The unpredictability of the climate will contribute to social and economic uncertainty. The planet in fifty years' time will look very different, and those of us who are still here to see in the next century will be living in a changed world.

Talking about what needs to be done is not exactly relaxing, either. It's not just about technological wizardry of electric vehicles and renewable energy. We do need to have those tricky conversations about eating less meat and dairy, and travelling differently or, indeed, not travelling at all. And we need to think about how these changes will come about – through protest, collective action, challenging power, electing different politicians, changing the law.

The philosopher Donna Haraway, whose writing is both brilliantly insightful and, at times, utterly impenetrable, coined a phrase which has lodged in my brain: 'staying with the trouble' (Haraway, 1988). I am not sure she quite defines it, so I am just going to say what I think it means. I think she's saying, if you identify something that needs dealing with – 'the trouble' – be brave enough to tackle it head on. There will be so many reasons not to. You will want to turn away, for the sake of an easy life. Others will want you to turn away, because what you say might conflict with their idea of an easy life. Staying with the trouble is, in short, troubling – and also brave, honest and constructive.

The Iraq War veteran Ray Scranton has written hauntingly about the psychological similarities of military service and climate change. He writes about how he got through his tour of duty in Iraq not by denying the dangers, but by confronting and examining them, and considering his own mortality. This, Scranton argues, is how we should be thinking of climate change:

> The choice is a clear one. We can continue acting as if tomorrow will be just like yesterday, growing less and

less prepared for each new disaster as it comes, and more and more desperately invested in a life we can't sustain. Or we can learn to see each day as the death of what came before, freeing ourselves to deal with whatever problems the present offers without attachment or fear. (Scranton, 2013)

I relate to this. I want to deal with the situation we are in, rather than hiding away. And yet I find myself reacting strongly to those like Scranton who focus unremittingly on the horrors of climate change. It feels like a self-fulfilling prophecy. It is a recipe for despair. This is why it is crucial to talk, also, about what can be done, and what has already been done: the growth in renewable energy, divestment from fossil fuels, the questions now being asked about diets and flying that just weren't part of the conversation a few years ago. The remarkable rise of climate change up the political agenda in 2018 and 2019 was a sign that rapid shifts are possible.

Take to the streets

Recent events have also shown the power of protest. The school strikes, involving nearly one and a half million students worldwide (Carrington, 2019), the Extinction Rebellion protests, and the mass lobby of the UK Parliament in June 2019, when twelve thousand people met outside Parliament to talk to over 300 MPs, as well as many, many local campaigns, have all had an impact. So protest if you can. From a politician's point of view, the less you think of yourself as a 'typical' protester, the more you are likely to be heard: showing your strength of feeling as a parent, as a nurse or teacher or business owner, as someone who doesn't normally take to the streets, is immensely powerful. Paradoxically, the less comfortable you feel about protesting, the more powerful a protester you will be.

In October 2019, I went to join the Extinction Rebellion protesters on the streets of London. It made me realise that the more different types of people joined in, the more powerful the protests would be. The conversations I had on the London streets, with grandparents, lawyers, school students, farmers and teachers, all of whom shared a common cause – to bring climate change to political attention – were truly life-affirming. I was moved by the quiet, peaceful determination of everyone I spoke to.

Think conversation, not monologue

I've suggested speaking out. Listening is just as important. I've noticed something interesting about meetings with politicians. I used to run a lot of them. I was responsible for gathering together environmental leaders to talk to government ministers. It was usually very one-sided. Environmentalists, who are, after all, known for their earnestness and commitment, would explain the problem, and then explain the solution. The minister would nod politely and drink coffee. We were placing endless demands upon them, but we weren't actually involving or engaging them at all.

Occasionally, though, it would turn into a conversation, and the chemistry changed. I remember an exchange with political advisers at the Treasury. Us environmentalists wanted to limit airport expansion. Fair enough, you might think. But the adviser then pointed out what, to him, was obvious: people want to fly, and he didn't want to be responsible for denying people that opportunity. He was adamant that flying shouldn't be something that only rich people could afford. Whether or not you agree with this, the point is: it became a dialogue. He knew what we wanted, and we knew why he found that a challenge. This frankness was very helpful – another example of 'staying with the trouble'. What followed was an honest and useful discussion about our views on whether, and under what circumstances, flying could be limited.

Acknowledging different views doesn't mean that you have to accept them, but it does mean that you have to take them seriously. It's surprising how rare this is. Yet it is a core principle of deliberation – listening to, and accepting, others' positions. Curiosity is a powerful weapon. In the experiment by Climate Outreach that I described in the section on 'speaking out', the advice that climate conversationalists found most useful was to respect the other person, ask questions, listen, and avoid blaming. It's a good recipe not just for climate conversations, but for all political ones.

Another aspect of 'staying with the trouble' is rejecting the stealth strategies discussed in Chapter Five – the idea that we can somehow tackle climate change without people noticing, through a focus on technology. On my daily trawl through Twitter, I am in turns amused and enraged by comments on new technologies that could 'save the planet'. A recent piece in *The Atlantic*, for example, promised that 'Climate change can be stopped by turning air into gasoline' (Meyer, 2018). There seem to be increasing numbers of very wealthy entrepreneurs putting forward climate 'solutions', such as Richard Branson's Earth Challenge, and Elon Musk's ultimate solution to problems on planet Earth: colonizing Mars. As Musk has said, "You back up your hard drive ... maybe we should back up life, too?" (quoted in Heath, 2015).

What strikes me about these proposals is that they offer evermore ambitious technological 'solutions' as an alternative to, or perhaps a distraction from, changes to social and economic systems. The idea of any significant change to the social order, distribution of wealth or powers of governments appears so preposterous that migration to Mars begins to seem more plausible than social reform and environmental governance on our own planet.

So in talking about, and acting on, climate, ask difficult questions about supposed planet-saving technologies. Focus instead on the ways in which we can live our lives better – and work to engage people, rather than bypassing them.

Change the rules

The New Economics Foundation, a charity that I am proud to be part of, works with communities across the UK who are finding new ways to tackle environmental and social issues. A lot of their work is immensely practical: setting up childcare cooperatives, campaigning for affordable housing, forming unions for people working on precarious zero-hours contracts. But it's also political. They want to show that change is possible, but they also want to show how the existing rules of the game can make it tough going. They want to change the rules. So they help people working on these practical projects to lobby the government, bringing this grassroots experience to bear.

There are so many amazing community initiatives on climate. Like community energy cooperatives, owned and run by local people, providing renewable energy and energy efficiency services. Like the People's Café in Kendal, where I live, who provide free or subsidized meals for people who need them, using food donated by local shops that would otherwise have been wasted. Like parents who arrange walking buses to encourage kids to walk to school.

The challenge, now, is to make all these community initiatives add up to something even greater – something that can change the rules. If you're involved in projects like this, thank about what rules need to change, and join campaigns to make it happen. Ask for help from national organizations, who would love to know about local projects trying to do things differently. Involve local MPs and councillors – and if they say how brilliant your project is, point out that it's their job to make this sort of thing easier. This interplay between practical action and pressure for change has huge political power.

Think about your own footprint, at home and work

I'm nearly at the end of my list of suggestions, and I have not yet told you to turn the lights off, to buy renewable energy, cut back on meat and dairy, stop flying, and insulate your house. Of course you should do this, and there are plenty of books and websites that can tell you what to do. Peter Kalmus, a US academic, has written a brilliant book about what each of us can do (Kalmus, 2017). Mike Berners-Lee's There is No Plan(et) B is another good one (Berners-Lee, 2019). But even with these personal actions, you should still think politically.

Your own carbon footprint is a drop in the global ocean. Every drop, like every vote, counts. It counts even more if you talk about it. What better way to talk about the need to reduce aviation than to say that you have restricted your own flying, for work and for holidays? Imagine how powerful it would be if everyone who campaigned for climate action – politicians, businesspeople, celebrities, everyone – made meaningful pledges about what they would do in their own lives. Could you be the person who prompts your organization to change?

There is a growing band of university researchers who have pledged to stop the wasteful amounts of flying that are currently a normal part of academic life. As a result, new options are opening up. International conferences have been run without air travel – like the 2018 'Displacements' anthropology conference, where online presentations were watched at different regional hubs (Pandian, 2018). When I write research grants, I factor in the time and money for train travel, not flights. I have also done some brilliant research using webinars rather than actual meetings. It's different, but it can work really well. On one memorable occasion, a workshop participant in California decided to show everyone joining from round the world his beautiful stripy knitted socks. I remember him waving his feet in front of his laptop camera.

It's not a case of all-or-nothing. My good friend Kate Rawles, an amazing adventurer and climate communicator, has set herself a budget of one flight every three years, and talks about this whenever she can. She says that people find it easier to relate to than stopping flying altogether (in rich countries, at least − it's always worth adding the caveat that most people in the world have never got on a plane). Similarly, I'm an occasional meat-eater − I don't think you have to choose between meat every day and a strict vegan diet. Do what you can − and tell people about it. There's research to show that it makes a difference. As I discussed in Chapter Four, people are heavily influenced by their social world. If people they respect have changed their behaviour significantly, this has an impact (Westlake, 2019).

Be kind to yourself

Lastly, as I discussed in Chapter Four, it is vitally important to acknowledge the personal psychological impact of the climate crisis. It is exhausting both to speculate about a climate-changed future, and to keep up the fight. Climate politics is emotional, and we should acknowledge this. I am very aware that I have privilege beyond measure, and that, for many people in many countries, climate impacts intersect with many other factors to make people's lives exceptionally difficult. But acknowledging that doesn't make it any less real for me. And so that brings me to my final suggestion: be kind to yourself. Acknowledge the emotional impact of climate, and find whatever coping mechanisms you can. Sometimes, protest and activism are good ways of coping. Sometimes, I just need to switch off and go running.

And finally

This moment, right now, is both the most exciting and the most unsettling phase of climate politics that I have known. Exciting, because climate is finally on the political agenda, with all its messy contradictions. Questions are being asked about where the money comes from, where the power lies, and what our society might look like as we move beyond fossil fuels. Unsettling, because wider political currents drag us into new conflicts, and threaten to turn people away from collective action, as the temptation to retreat into personal protection intensifies. And unsettling because the earth is entering climate conditions as yet unknown to the human species.

There is no hiding the fact that I'm nervous about the years ahead. My children, now teenagers, are growing up in a world that is more troubled and less certain than ever. And yet I find that I can't retreat into a narrative of grief, loss and pessimism. Neither have I lost my faith in people to understand, accept and act on the unfolding crisis. I know one thing for sure: that it is both morally wrong and logically flawed to bypass democracy, impose solutions and shut down debate. The more we push the politics of climate into the harsh light of the public sphere, and expose the contradictions of high-carbon societies and political life, the more likely it is that we will be able to move beyond them. It's about staying with the trouble.

Staying with the trouble also means acknowledging that the future is yet to be shaped. Through the often frustrating but also joyous world of social media, I have got to know some amazing people, many of whom I have never met in person. Meg Ruttan Walker, a teacher, activist and mother in Canada, is one of those. She writes of the dangers of mythologizing the future, either by telling ourselves a story that we can continue as we are or the equally dangerous but beguiling story that the catastrophists tell: it's all too late. In her own, beautiful words:

> The future isn't a prophecy. It's a living document that is being written, erased, rewritten, and written over right now. By too many hands to count. And, consequently, skipping to the end seems foolish, and impossible, to me. (Ruttan Walker, 2019)

A proper political response to the climate crisis means acknowledging this deep uncertainty, and expecting people to respond thoughtfully and with dignity. It requires deeper engagement, and a better acknowledgement that climate change is not just a scientific or technical issue, but a deeply social one. It is about how people act collectively to shape their society, within the confines of this increasingly changeable planet that is home to us all.

References

Chapter One

Amladi, D., 2017. Raising a family in a time of extreme hunger [WWW Document]. Oxfam America. URL /explore/stories/raising-a-family-in-a-time-of-extreme-hunger/ (accessed 25.10.19).

Brown, M., 2015. Gormley climate change artwork shown for first time. *The Guardian* 6 March https://www.theguardian.com/environment/2015/mar/06/gormley-climate-change-artwork-connection-shown-for-first-time-in-guardian

Buck, G., 2019. Power To The People. Green Alliance, London.

Childs, S., 2004. A feminised style of politics? Women MPs in the House of Commons. British Journal of Politics and International Relations 6, 3–19.

Crewe, E., 2015. The House of Commons: An anthropology of MPs at work. London: Bloomsbury.

Demeritt, D., 2001. The construction of global warming and the politics of science. Annals of the Association of American Geographers 91, 307–37. https://doi.org/10.1111/0004-5608.00245

Díaz, S., Settele, J., Brondízio, E., Ngo, H.T., Guèze, M., Agard, J., Arneth, A., Balvanera, P., Brauman, K., Watson, R.T., Baste, I.A., Larigauderie, A., Leadley, P., Pascual, U., Baptiste, B., Demissew, S., Dziba, L., Erpul, G., Fazel, A., Fischer, M., María, A., Karki, M., Mathur, V., Pataridze, T., Pinto, I.S., Stenseke, M., Török, K., Vilá, B., 2019. Summary for policymakers of the global assessment report on biodiversity and ecosystem services of the Intergovernmental Science-Policy Platform on Biodiversity and Ecosystem Services. IPBES.

Fenno Jr, R., 1977. US House Members in their constituencies: an exploration. The American Political Science Review 71, 883–917.

Fielding, K.S., Head, B.W., Laffan, W., Western, M., Hoegh-Guldberg, O., 2012. Australian politicians' beliefs about climate change: political partisanship and political ideology. Environmental Politics 21, 712–33. https://doi.org/10.1080/09644016.2012.698887

Haraway, D., 1988. Situated knowledges: the science question in feminism and the privilege of partial perspective. Feminist Studies 14, 575–99. https://doi.org/10.2307/3178066

Hickman, L., 2010. James Lovelock: "Fudging data is a sin against science". The Guardian 3 March 2010 https://www.theguardian.com/environment/2010/mar/29/james-lovelock

Intergovernmental Panel on Climate Change, 2018. Global warming of 1.5°C. http://www.ipcc.ch/report/sr15/

Jackson, T., 2017. Prosperity without Growth: Foundations for the economy of tomorrow. Second edition. Routledge, Taylor & Francis Group, London; New York.

Jasanoff, S., 2010. A new climate for society. Theory Culture Society 27, 233–53. https://doi.org/10.1177/0263276409361497

Lovenduski, J., 2012. Prime Minister's questions as political ritual. British Politics 7, 314–40.

Malley, R., 2012. The institutionalisation of gendered norms and the substantive representation of women in Westminster and the Scottish parliament (PhD thesis). University of Bristol.

McKay, J., 2011. 'Having it all?' Women MPs and motherhood in Germany and the UK. Parliamentary Affairs 64, 714–36. https://doi.org/10.1093/pa/gsr001

McNeil, M., 2013. Between a rock and a hard place: the deficit model, the diffusion model and publics in STS. Science as Culture 22, 589–608. https://doi.org/10.1080/14636778.2013.764068

Norton, P., 2012. Parliament and citizens in the United Kingdom. The Journal of Legislative Studies 18, 403–18. https://doi.org/10.1080/13572334.2012.706053

Oxfam, 2015. Extreme carbon inequality: why the Paris climate deal must put the poorest, lowest emitting and most vulnerable people first. https://doi.org/10.1163/2210-7975_HRD-9824-2015053

REFERENCES

Riessman, C.K., 2008. Narrative Methods for the Human Sciences. Sage Publications, Los Angeles.

Rockström, J., Steffen, W., Noone, K., Persson, Asa, Chapin, F.S., Lambin, E.F., Lenton, T.M., Scheffer, M., Folke, C., Schellnhuber, H.J., and others, 2009. A safe operating space for humanity. Nature 461, 472–5.

Schleussner, C.-F., Deryng, D., Müller, C., Elliott, J., Saeed, F., Folberth, C., Liu, W., Wang, X., Pugh, T.A.M., Thiery, W., Seneviratne, S.I., Rogelj, J., 2018. Crop productivity changes in 1.5°C and 2°C worlds under climate sensitivity uncertainty. Environmental Research Letters 13, 064007. https://doi.org/10.1088/1748-9326/aab63b

Stern, N. (2018) We must reduce greenhouse gas emissions to net zero or face more floods. *The Guardian*, Monday 8 October.

Wallace-Wells, D., 2019. The Uninhabitable Earth: Life after warming. First edition. Tim Duggan Books, New York.

Weingartner, P., 2018. A harrowing personal account of how the camp fire devastated paradise [WWW Document]. Jefferson Public Radio. URL https://www.ijpr.org/post/harrowing-personal-account-how-camp-fire-devastated-paradise (accessed 25.10.19).

Willis, R., 2017. Taming the climate? Corpus analysis of politicians' speech on climate change. Environmental Politics 26, 212–31. https://doi.org/10.1080/09644016.2016.1274504

Willis, R., 2018a. How Members of Parliament understand and respond to climate change. The Sociological Review 66, 475–91. https://doi.org/10.1177/0038026117731658

Willis, R., 2018b. Constructing a 'representative claim' for action on climate change: evidence from interviews with politicians. Political Studies 003232171775372. https://doi.org/10.1177/0032321717753723

Wilsdon, J., Willis, R., 2004. See-through Science: Why public engagement needs to move upstream. London: Demos.

World Bank, 2019. CO_2 emissions (metric tonnes per capita) [WWW Document]. URL https://data.worldbank.org/indicator/EN.ATM.CO2E.PC?most_recent_value_desc=false (accessed 25.10.19).

Wynne, B., 2010. Strange weather, again: Climate science as political art. Theory Culture Society 27, 289–305. https://doi.org/10.1177/0263276410361499

Chapter Two

Ainsley, C. (2018). The new working class: how to win hearts, minds and votes. Bristol: Policy Press.

BBC News, 2009. Maldives cabinet makes a splash. BBC. 17 October, http://news.bbc.co.uk/1/hi/8311838.stm

Castree, N., 2014. The Anthropocene and geography I: the back story. Geography Compass 8, 436–49. https://doi.org/10.1111/gec3.12141

Climate Action Tracker, 2019a. USA | Climate Action Tracker [WWW Document]. Climate Action Tracker. URL https://climateactiontracker.org/countries/usa/ (accessed 25.10.19).

Climate Action Tracker, 2019b. Ethiopia | Climate Action Tracker [WWW Document]. Climate Action Tracker. URL https://climateactiontracker.org/countries/ethiopia/ (accessed 25.10.19).

Dale, D., 2019. Every false claim Donald Trump has made as president [WWW Document]. Toronto Star. URL https://projects.thestar.com/donald-trump-fact-check/ (accessed 25.10.19).

Darwall, R. (2017). Will Trump stand up to the world on climate change policy? National Review, 22 February 2017.

Dobson, A., 2010. Democracy and nature: speaking and listening. Political Studies 58, 752–68. https://doi.org/10.1111/j.1467-9248.2010.00843.x

Fawcett, A.A., et al. (2015). Can Paris pledges avert severe climate change? Science 350, 1168–1169.

Foster, C., Frieden, J., 2017. Crisis of trust: socio-economic determinants of Europeans' confidence in government. European Union Politics 18, 511–35. https://doi.org/10.1177/1465116517723499

Fuller, R.B., 1969. Operating Manual For Spaceship Earth. Southern Illinois University Press.

Funk, C., Kennedy, B., 2016. The Politics of Climate. Washington: Pew Research Center.

Hajer, M., Nilsson, M., Raworth, K., Bakker, P., Berkhout, F., de Boer, Y., Rockström, J., Ludwig, K., Kok, M., 2015. Beyond cockpit-ism: four insights to enhance the transformative potential of the Sustainable Development Goals. Sustainability 7, 1651–60. https://doi.org/10.3390/su7021651

Guiney, A., 2018. Why I carried this sign to protest government action on climate change, Women's Agenda, December 2 https://womensagenda.com.au/latest/climate-change-activist-alicia-gueney/

IRENA, 2018. Renewable energy jobs reach 10.3 million worldwide in 2017 [WWW Document]. /newsroom/pressreleases/2018/May/Renewable-Energy-Jobs-Reach-10-Million-Worldwide-in-2017. URL /newsroom/pressreleases/2018/May/Renewable-Energy-Jobs-Reach-10-Million-Worldwide-in-2017 (accessed 6.9.18).

Jasanoff, S., 2010. A new climate for society. Theory Culture Society 27, 233–53. https://doi.org/10.1177/0263276409361497

Lockwood, M., 2013. The political sustainability of climate policy: the case of the UK Climate Change Act. Global Environmental Change 23, 1339–48. https://doi.org/10.1016/j.gloenvcha.2013.07.001

Mandel, K., 2016. Revealed: the climate science deniers behind the Brexit campaign. openDemocracy. https://www.opendemocracy.net/uk/kyla-mandel/revealed-climate-science-deniers-behind-brexit-campaign

Memmott, M., 2013. "Stop this madness," tearful Filipino pleads at climate talks [WWW Document]. NPR.org. URL https://www.npr.org/sections/thetwo-way/2013/11/11/244533348/stop-this-madness-tearful-filipino-pleads-at-climate-talks (accessed 25.10.19).

Reuben, A., Barnes, P., 2016. Checking the Vote Leave leaflet. BBC News, 4 April https://www.bbc.com/news/uk-politics-eu-referendum-36014941

Rockström, J., Steffen, W., Noone, K., Persson, Asa, Chapin, F.S., Lambin, E.F., Lenton, T.M., Scheffer, M., Folke, C., Schellnhuber, H.J., and others, 2009. A safe operating space for humanity. Nature 461, 472–5.

Scoones, I., Leach, M., Newell, P., 2015. The Politics of Green Transformations. Taylor and Francis, Florence.

Shahar, D.C., 2015. Rejecting eco-authoritarianism, again [WWW Document]. https://doi.org/info:doi/10.3197/096327114X13947900181996

Stirling, A., 2015. 'Reigning back' the Anthropocene is hard – but Earth's worth it [WWW Document]. STEPS Centre. URL https://steps-centre.org/blog/reigning-back-the-anthropocene-is-hard-but-earths-worth-it/ (accessed 25.10.19).

Strahan, S.E., Douglass, A.R., 2018. Decline in Antarctic ozone depletion and lower stratospheric chlorine determined from aura microwave limb sounder observations. Geophysical Research Letters 45, 382–90. https://doi.org/10.1002/2017GL074830

Thatcher, M., 1989. Speech to the UN General Assembly, November 8 1989.

Timperley, J., 2019. The Carbon Brief profile: India [WWW Document]. URL https://www.carbonbrief.org/the-carbon-brief-profile-india (accessed 25.10.19).

Tyfield, D., 2018. Liberalism 2.0 and the Rise of China: global crisis, innovation and urban mobility, Routledge Advances in Sociology. Routledge/Taylor & Francis Group, New York; London.

United Nations (UN), 1992. United Nations Framework Convention on Climate Change. https://unfccc.int/files/essential_background/background_publications_htmlpdf/application/pdf/conveng.pdf

Zhu, Y., 2011. 'Performance legitimacy' and China's political adaptation strategy. Journal of Chinese Political Science 16, 123–40. https://doi.org/10.1007/s11366-011-9140-8

Chapter Three

Alston, P., 2019. Promotion and protection of all human rights, civil, political, economic, social and cultural rights, including the right to development. https://doi.org/10.1163/2210-7975_HRD-9970-2016149

Berners-Lee, M., 2019. There is No Plan(et) B: A handbook for the make or break years. Cambridge University Press, Cambridge; New York, NY.

Berners-Lee, M., Clark, D., 2013. The Burning Question: We can't burn half the world's oil, coal and gas; so how do we quit? First edition. Profile Books, London.

Capellán-Pérez, I., de Castro, C., Miguel González, L.J., 2019. Dynamic Energy Return on Energy Investment (EROI) and material requirements in scenarios of global transition to renewable energies. Energy Strategy Reviews 26, 100399. https://doi.org/10.1016/j.esr.2019.100399

Carbon Tracker, 2011. Unburnable Carbon – Are the world's financial markets carrying a carbon bubble? Carbon Tracker Initiative. https://www.banktrack.org/download/unburnable_carbon/unburnablecarbonfullrev2.pdf

Catton, W., 2011. Speech to Annual Conference of the Association for the Study of Peak Oil and Gas. 3 November, reported in https://grist.org/article/2011-11-04-each-american-consumes-as-much-energy-as-40-ton-dinosaur/

Climate Tracker, 2018. Over 25 fossil fuel representatives part of country delegations at COP24. Climate Tracker. URL http://climatetracker.org/over-25-fossil-fuel-representatives-part-of-country-delegations-at-cop24/ (accessed 22.10.19).

Crenson, M.A., 1971. The Un-politics of Air Pollution: A study of non-decisionmaking in the cities. Johns Hopkins Press, Baltimore.

Department for Transport, 2019. 2018 National Travel Survey. UK.

Dobson, A., 2013. Political theory in a closed world: Reflections on William Ophuls, liberalism and abundance. Environmental Values 22, 241–59. https://doi.org/10.3197/096327113X13581561725275

Exxon, 1982. 1982 memo to Exxon management about CO_2 greenhouse effect. Climate Files. URL http://www.climatefiles.com/exxonmobil/1982-memo-to-exxon-management-about-co2-greenhouse-effect/ (accessed 22.10.19).

Garside, M., 2019. ExxonMobil Stats & Facts [WWW Document]. www.statista.com. URL https://www.statista.com/topics/1109/exxonmobil/ (accessed 22.10.19).

Hickman, L., 2018. The Carbon Brief interview: Saudi Arabia's Ayman Shasly [WWW Document]. Carbon Brief. URL https://www.carbonbrief.org/the-carbon-brief-interview-saudi-arabias-ayman-shasly (accessed 23.10.19).

International Energy Agency, 2018a. Global Energy & CO_2 Status Report 2018. URL https://www.iea.org/geco/

International Energy Agency, 2018b. Population without access to electricity falls below 1 billion [WWW Document]. URL https://www.iea.org/newsroom/news/2018/october/population-without-access-to-electricity-falls-below-1-billion.html (accessed 22.10.19).

Kottasová, I., 2019. Oil helped Norway build up a $1 trillion fund. Now it's dumping oil stocks [WWW Document]. CNN. URL https://www.cnn.com/2019/03/08/investing/norway-fund-oil-stocks/index.html (accessed 22.10.19).

Macalister, T., 2000. BP rebrands on a global scale. The Guardian, 25 July. URL https://www.theguardian.com/business/2000/jul/25/bp (accessed 22.10.19).

Ministry of Defence, 2014. Global Strategic Trends out to 2040 [WWW Document]. URL https://www.gov.uk/government/uploads/system/uploads/attachment_data/file/33717/GST4_v9_Feb10.pdf (accessed 19.2.15).

Mitchell, T., 2011. Carbon Democracy: Political power in the age of oil. Verso, London; New York.

Murray, J., 2019. Is it Beyond Petroleum for real this time? Business Green. URL https://www.businessgreen.com/bg/blog-post/3070343/is-it-beyond-petroleum-for-real-this-time (accessed 22.10.19).

Nuccitelli, D., 2015. Two-faced Exxon: the misinformation campaign against its own scientists. The Guardian. 25 November https://www.theguardian.com/environment/climate-consensus-97-per-cent/2015/nov/25/two-faced-exxon-the-misinformation-campaign-against-its-own-scientists

Ophuls, W., 1992. Ecology and the Politics of Scarcity Revisited: The unraveling of the American dream. W.H. Freeman, New York.

Oreskes, N., Conway, E.M., 2012. Merchants of Doubt: How a handful of scientists obscured the truth on issues from tobacco smoke to global warming. Bloomsbury, London.

Parveen, N., 2019. National Theatre to end Shell membership from next year. The Guardian. 4 October https://www.theguardian.com/stage/2019/oct/04/national-theatre-to-end-shell-sponsorship-deal-from-next-year

Peters, A., 2018. Exxon thinks it can create biofuel from algae at massive scale [WWW Document]. Fast Company. URL https://www.fastcompany.com/40539606/exxon-thinks-it-can-create-biofuel-from-algae-at-massive-scale (accessed 22.10.19).

Shepherd, A., Gilbert, L., Muir, A.S., Konrad, H., McMillan, M., Slater, T., Briggs, K.H., Sundal, A.V., Hogg, A.E., Engdahl, M.E., 2019. Trends in Antarctic ice sheet elevation and mass. Geophysical Research Letters 46, 8174–83. https://doi.org/10.1029/2019GL082182

Teirstein, Z., 2018. What's with Exxon's big algae push? Grist. URL https://grist.org/article/whats-with-exxons-big-algae-push/ (accessed 22.10.19).

Urry, J., 2011. Climate Change and Society. Polity Press, Cambridge, U.K.; Malden, Mass.

Wainwright, J., Mann, G., 2018. Climate Leviathan: A political theory of our planetary future. Verso, London; New York.

Willis, R., Eyre, N., 2011. Demanding Less: Why we need a new politics of energy. Green Alliance, London.

World Bank, 2014. Energy use (kg of oil equivalent per capita) [WWW Document]. URL https://data.worldbank.org/indicator/EG.USE.PCAP.KG.OE (accessed 10.22.19).

Chapter Four

Barasi, L., 2017. The Climate Majority: Apathy and action in an age of nationalism. New Internationalist, Oxford.

Buck, G., 2019. Power To The People. Green Alliance, London.

Cecil, N., 2019. Climate change fears gripping Britain: poll reveals 85% are worried about warming [WWW Document]. Evening Standard. URL https://www.standard.co.uk/news/uk/climate-change-fears-gripping-britain-poll-reveals-85-are-worried-about-warming-the-highest-figure-a4218251.html (accessed 23.10.19).

Ebrey, R., 2019. Is Twitter indicating a change in MPs' views on climate change? Inside Track. URL https://greenallianceblog.org.uk/2019/10/01/is-twitter-indicating-a-change-in-mps-views-on-climate-change/ (accessed 23.10.19).

Hallam, R., 2019. BBC World Service HARDtalk. 18 August https://www.bbc.co.uk/programmes/w3csy93l

Head, L., Harada, T., 2017. Keeping the heart a long way from the brain: the emotional labour of climate scientists. Emotion, Space and Society, On trauma, geography, and mobility: Towards geographies of trauma 24, 34–41. https://doi.org/10.1016/j.emospa.2017.07.005

Kingsnorth, P., Hine, D., Dark Mountain Project, 2014. Uncivilisation: the Dark Mountain manifesto. https://dark-mountain.net/about/manifesto/

Lawler, S., 2014. Identity: Sociological perspectives. Second edition. Polity Press, Cambridge.

Lewis, G.B., Palm, R., Feng, B., 2019. Cross-national variation in determinants of climate change concern. Environmental Politics 28, 793–821. https://doi.org/10.1080/09644016.2018.1512261

Lucas, C., 2015. Honourable friends? Parliament and the fight for change. Portobello Books, London.

McCright, A.M., Dunlap, R.E., 2011. Cool dudes: The denial of climate change among conservative white males in the United States. Global Environmental Change 21, 1163–72. https://doi.org/10.1016/j.gloenvcha.2011.06.003

Norgaard, K.M., 2006. "We don't really want to know": environmental justice and socially organized denial of global warming in Norway. Organization Environment 19, 347–70. https://doi.org/10.1177/1086026606292571

Petersen, M., 2018. The Impact of Climate Migration. MSc thesis.

Poushter, J., Huang, C., 2019. Climate change still seen as top global threat, but cyberattacks rising concern. Pew Research Center's Global Attitudes Project. URL https://www.pewresearch.org/global/2019/02/10/climate-change-still-seen-as-the-top-global-threat-but-cyberattacks-a-rising-concern/ (accessed 23.10.19).

Puwar, N., 2004. Space Invaders: Race, gender and bodies out of place. Berg, London.

WHO, 2014. Quantitative risk assessment of the effects of climate change on selected causes of death, 2030s and 2050s. World Health Organization. https://www.who.int/globalchange/publications/quantitative-risk-assessment/en/

Willis, R., 2017. Taming the climate? Corpus analysis of politicians' speech on climate change. Environmental Politics 26, 212–31. https://doi.org/10.1080/09644016.2016.1274504

Chapter Five

Berners-Lee, M., Clark, D., 2013. The Burning Question: We can't burn half the world's oil, coal and gas; so how do we quit? First edition. Profile Books, London.

Lake District National Park Authority, 2018. Carbon budget for the Lake District [WWW Document]. Lake District National Park. URL https://www.lakedistrict.gov.uk/caringfor/lake-district-national-park-partnership/carbonbudget (accessed 25.10.19).

McGlade, C., Ekins, P., 2015. The geographical distribution of fossil fuels unused when limiting global warming to 2°C. Nature 517, 187–90. https://doi.org/10.1038/nature14016

Willis, R., 2011. Green Economy Council: a call to action, or paralysis by analysis? [WWW Document] URL http://www.rebeccawillis.co.uk/16/02/2011/green-economy-council-a-call-to-action-or-paralysis-by-analysis (accessed 14.1.15).

Willis, R., Eyre, N., 2011. Demanding Less: Why we need a new politics of energy. Green Alliance, London.

Wiseman, J., Edwards, T., Luckins, K., 2013. Post carbon pathways: a meta-analysis of 18 large-scale post carbon economy transition strategies. Environmental Innovation and Societal Transitions 8, 76–93. https://doi.org/10.1016/j.eist.2013.04.001

Chapter Six

Ainsley, C., 2018. The New Working Class: How to win hearts, minds and votes. Policy Press, Bristol.

Blondiaux, L., 2019. Convention citoyenne pour le climat : "Il est impératif d'inventer de nouvelles formes de démocraties". FranceInter, Paris.

Broockman, D.E., Skovron, C., 2013. What Politicians Believe About Their Constituents: Asymmetric Misperceptions and Prospects for Constituency Control 51 Working Paper prepared for presentation at 'Political Representation: Fifty Years After Miller and Stokes,' Vanderbilt University, March 1-2, 2013. URL https://www.vanderbilt.edu/csdi/miller-stokes/08_MillerStokes_BroockmanSkovron.pdf.

Conservative Environment Network, 2019. The CEN Manifesto. https://www.cen.uk.com/manifesto

Curato, N., Dryzek, J.S., Ercan, S.A., Hendriks, C.M., Niemeyer, S., 2017. Twelve key findings in deliberative democracy research [WWW Document]. American Academy of Arts & Sciences. URL https://www.amacad.org/publication/twelve-key-findings-deliberative-democracy-research (accessed 22.8.19).

Darby, M., 2019. Ireland to 'nudge' its way to net zero emissions by 2050 [WWW Document]. Climate Home News. URL https://www.climatechangenews.com/2019/06/18/ireland-nudge-way-net-zero-emissions-2050/ (accessed 24.10.19).

Dryzek, J.S., Niemeyer, S., 2008. Discursive representation. The American Political Science Review 102, 481–93.

Evans, A., 2019. The myth gap: how to navigate a world of 'post-truth' politics. URL https://www.collectivepsychology.org/myth-gap-post-truth/ (accessed 24.10.19). Collective Psychology.

Heglar, M.A., 2019. But the greatest of these is love [WWW Document]. Medium. URL https://medium.com/@maryheglar/but-the-greatest-of-these-is-love-4b7aad06e18c (accessed 24.10.19).

Lehr, R.L., Guild, W., Thomas, D.L., Swezey, B.G., 2003. Listening to customers: how deliberative polling helped build 1,000 MW of new renewable energy projects in Texas (No. NREL/TP-620-33177, 15003900). https://doi.org/10.2172/15003900

Milanović, B., 2016. Global Inequality: A new approach for the age of globalization. The Belknap Press of Harvard University Press, Cambridge, Massachusetts.

Saward, M., 2010. The Representative Claim. Oxford University Press, Oxford.

The Citizens' Assembly, 2018. Third Report and Recommendations of the Citizens' Assembly: How the state can make Ireland a leader in tackling climate change. 18 April https://www.citizensassembly.ie/en/How-the-State-can-make-Ireland-a-leader-in-tackling-climate-change/Final-Report-on-how-the-State-can-make-Ireland-a-leader-in-tackling-climate-change/Climate-Change-Report-Final.pdf

Willis, R., 2017. Taming the climate? Corpus analysis of politicians' speech on climate change. Environmental Politics 26, 212–31. https://doi.org/10.1080/09644016.2016.1274504

Chapter Seven

Armstrong, C., 2019. Decarbonisation and world poverty: a just transition for fossil fuel exporting countries? Political Studies 003232171986821. https://doi.org/10.1177/0032321719868214

Buck, G., 2019. Power To The People. Green Alliance, London.

Caulkin, S., Collins, J., 2003. The Private Life of Public Affairs. Green Alliance, London.

Climate Emergency, 2019. Climate Emergency Website. URL https://www.climateemergency.uk/ (accessed 25.10.19).

Cooper, A., 2018. Regionally and Locally Determined Contributions. Committee of the Regions. November 2018 https://cor.europa.eu/en/news/Documents/3894-leaflet-Cooper-v3-LR.PDF

Denyer, 2018. Tweet by Carla Denyer [WWW Document]. Twitter. URL https://twitter.com/carla_denyer/status/1066665378590453760?ref_url=https%3a%2f%2fclimateemergencydeclaration.org%2funited-kingdom-bristol-city-council-declares-a-climate-emergency%2f (accessed 25.10.19).

HM Government, 2009. The UK Low Carbon Transition Plan: National strategy for climate and energy. Stationery Office, London.

Jackson, T., 2017. Prosperity without Growth: Foundations for the economy of tomorrow, Second Edition. ed. Routledge, Taylor & Francis Group, London; New York.

Johnson, B., 2019. Dear Extinction Rebellion: your aims are worthy, but take your pink boat to China instead. *The Telegraph*, 21 April 2019.

Klein, N., 2015. This Changes Everything: Capitalism vs. the climate. Penguin Books, London.

Machin, A., 2013. Negotiating Climate Change: Radical democracy and the illusion of consensus. Zed Books, London.

McGlade, C., Ekins, P., 2015. The geographical distribution of fossil fuels unused when limiting global warming to 2°C. Nature 517, 187–90. https://doi.org/10.1038/nature14016

McLaren, D.P., Tyfield, D.P., Willis, R., Szerszynski, B., Markusson, N.O., 2019. Beyond 'Net-Zero': A case for separate targets for emissions reduction and negative emissions. Frontiers in Climate 1. https://doi.org/10.3389/fclim.2019.00004

Mouffe, C., 2000. The Democratic Paradox. Verso, London; New York.

Newell, P., Simms, A., 2019. Towards a fossil fuel non-proliferation treaty. Climate Policy 0, 1–12. https://doi.org/10.1080/14693062.2019.1636759

Schatelek, L., Nakhooda, S., Watson, C., 2016. Climate Funds Update: The Green Climate Fund. Henrich Böll Stiftung. URL https://www.odi.org/sites/odi.org.uk/files/resource-documents/11050.pdf

Scottish Government, 2019. The Government's Programme for Scotland 2019-20 162. URL https://www.gov.scot/publications/protecting-scotlands-future-governments-programme-scotland-2019-20/

Taylor, M., Watts, J., 2019. Revealed: the 20 firms behind a third of all carbon emissions. The Guardian. 10 September https://www.theguardian.com/environment/2019/oct/09/revealed-20-firms-third-carbon-emissions

Tong, D., Zhang, Q., Zheng, Y., Caldeira, K., Shearer, C., Hong, C., Qin, Y., Davis, S.J., 2019. Committed emissions from existing energy infrastructure jeopardize 1.5°C climate target. Nature 572, 373–7. https://doi.org/10.1038/s41586-019-1364-3

Vaughan, A., 2017. Energy bills: what's the difference between Tory cap and Miliband freeze? The Guardian. 23 April https://www.theguardian.com/money/2017/apr/23/energy-prices-tory-cap-miliband-freeze

Wallace, A.A., Irvine, K.N., Wright, A.J., Fleming, P.D., 2010. Public attitudes to personal carbon allowances: findings from a mixed-method study. Climate Policy 385–409.

Wappelhorst, S., 2018. Using vehicle taxation policy to lower transport emissions: An overview for passenger cars in Europe. The International Council on Clean Transportation, December. URL: https://theicct.org/sites/default/files/publications/EU_vehicle_taxation_Report_20181214_0.pdf

Willis, R., 2017. Taming the climate? Corpus analysis of politicians' speech on climate change. Environmental Politics 26, 212–31. https://doi.org/10.1080/09644016.2016.1274504

Willis, R., Mitchell, C., Hoggett, R., Britton, J., 2019a. Enabling the transformation of the energy system: Recommendations from IGov. University of Exeter. http://projects.exeter.ac.uk/igov/enabling-the-transformation-of-the-energy-system/

Willis, R., Mitchell, C., Hoggett, R., Britton, J., Poulter, H., Pownall, T., Lowes, R., 2019b. Getting energy governance right: Lessons from IGov. University of Exeter. https://projects.exeter.ac.uk/igov/getting-energy-governance-right-lessons-from-igov/

Chapter Eight

Berners-Lee, M., 2019. There is No Plan(et) B: A handbook for the make or break years. Cambridge University Press, Cambridge; New York, NY.

Carrington, D., 2019. School climate strikes: 1.4 million people took part, say campaigners. The Guardian. 19 March https://www.theguardian.com/environment/2019/mar/19/school-climate-strikes-more-than-1-million-took-part-say-campaigners-greta-thunberg

Haraway, D., 1988. Situated knowledges: the science question in feminism and the privilege of partial perspective. Feminist Studies 14, 575–99. https://doi.org/10.2307/3178066

Heath, C., 2015. Elon Musk is ready to conquer Mars. GQ. December https://www.gq.com/story/elon-musk-mars-spacex-tesla-interview

Kalmus, P., 2017. Being the Change: Live well and spark a climate revolution. New Society Publishers, Gabriola Island, BC.

Meyer, R., 2018. Climate change can be reversed by turning air into gasoline [WWW Document]. The Atlantic. URL https://www.theatlantic.com/science/archive/2018/06/its-possible-to-reverse-climate-change-suggests-major-new-study/562289/ (accessed 7.9.18).

Pandian, A., 2018. Reflections on #displace18 [WWW Document]. Society for Cultural Anthropology. URL https://culanth.org/about/about-the-society/announcements/reflections-on-displace18 (accessed 25.10.19).

Ruttan Walker, M., 2019. Mythologizing the future. A Variety of Meltdowns. URL https://avarietyofmeltdowns.wordpress.com/2019/08/21/mythologizing-the-future/ (accessed 25.10.19).

Scranton, R., 2013. Learning how to die in the Anthropocene [WWW Document]. The New York Times. URL http://opinionator.blogs.nytimes.com/2013/11/10/learning-how-to-die-in-the-anthropocene/ (accessed 19.2.15).

Webster, R., 2019a. The Talking Climate Handbook: How to have conversations about climate change in your daily life. Climate Outreach.

Webster, R., 2019b. Private communication.

Westlake, S., 2019. Climate change: yes, your individual action does make a difference [WWW Document]. The Conversation. URL http://theconversation.com/climate-change-yes-your-individual-action-does-make-a-difference-115169 (accessed 25.10.19).

Index